WILD RIVER, TIMELESS CANYONS

Balduin Möllhausen als Trapper.

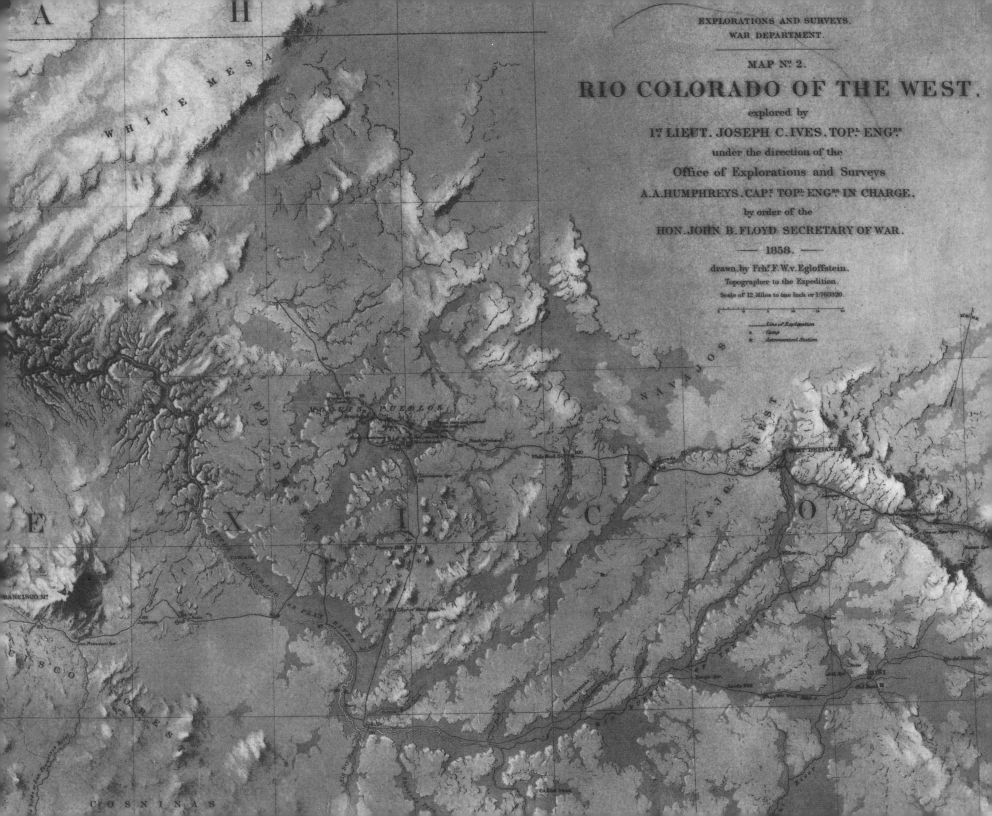

EXPLORATIONS AND SURVEYS.
WAR DEPARTMENT.

MAP Nº 2.

RIO COLORADO OF THE WEST.

explored by

1ST LIEUT. JOSEPH C. IVES, TOPL ENGRS

under the direction of the

Office of Explorations and Surveys

A. A. HUMPHREYS, CAPT TOPL ENGRS IN CHARGE.,

by order of the

HON. JOHN B. FLOYD SECRETARY OF WAR.

——— 1858. ———

drawn by Frhr F. W. v. Egloffstein.
Topographer to the Expedition.
Scale of 12 Miles to one Inch or 1:760320.

_____ Line of Exploration
⌂ Camp
✳ Astronomical Station

WILD RIVER, TIMELESS CANYONS

Balduin Möllhausen's
Watercolors of the Colorado

Ben W. Huseman

AMON CARTER MUSEUM
Fort Worth, Texas

Distributed by
UNIVERSITY OF ARIZONA PRESS
Tucson, Arizona

©1995, Amon Carter Museum

Designed by Tom Dawson, Fort Worth, Texas
Printed by South China Printing Co. (1988) Ltd., Hong Kong

ISBN 0-88360-084-6 (hardcover)

LIBRARY OF CONGRESS CATALOGING-IN-PUBLICATION DATA
Amon Carter Museum of Western Art.
     Wild river, timeless canyons: Balduin Möllhausen's watercolors of
the Colorado / Ben W. Huseman.
        p. cm.
     A catalog of forty-six watercolors in the Amon Carter Museum.
     Includes bibliographical references and index.
     ISBN 0-88360-084-6
     1. Möllhausen, Balduin. 1825-1905—Catalogs. 2. Colorado River
(Colo.-Mexico) in art—Catalogs. 3. Watercolor painting—Texas—Fort
Worth—Catalogs. 4. Amon Carter Museum of Western Art—Catalogs.
I. Huseman, Ben W. II. Title.
ND1954.M65A4 1995
759.3—dc20   95-35223
                                               CIP

The Amon Carter Museum was established through the generosity of Amon G. Carter, Sr. (1879-1955) to house his collection of paintings and sculpture by Frederic Remington and Charles M. Russell; to collect, preserve, and exhibit the finest examples of American art; and to serve an educational role through exhibitions, publications, and programs devoted to the study of American art.

Frontispiece: Möllhausen in his trapper's outfit, steel engraving in *Die Gartenlaube* 10, no. 29 (1862), p. 453, photograph courtesy of Perry-Castaneda Library, University of Texas, Austin

Contents

FOREWORD

During the middle decades of the nineteenth century, scientists viewed the great expanse of the American West as a wondrous laboratory for studying climatology, geography, geology, and the relationship of plants and animals to their physical surroundings. Under the seminal influence of the German explorer and scientist Alexander von Humboldt (1769-1859), new discoveries and observations were increasingly conveyed in visual terms, for greater ease of comprehension; the field of geologic cartography, for example, was entirely transformed in this period. Baron von Humboldt's protégé, the German artist and writer Heinrich Balduin Möllhausen (1825-1905), was an active participant in these developments. During the 1850s he made three trips to America, two of them to accompany United States government expeditions exploring the Trans-Mississippi West. The most memorable of these was perhaps the last: an expedition led by Lieutenant Joseph C. Ives of the Corps of Topographical Engineers, who set out in 1857 to explore the Colorado River and its magnificent surroundings. This publication chronicles that grand and historic adventure, through Möllhausen's vivid words and watercolors.

This project began more than seven years ago, when the Museum acquired a long-lost set of forty-six watercolors that Möllhausen had created from his field sketches of the Ives expedition. We owe thanks first of all to Harry L. Stern, a long-time friend of the Museum, for bringing these watercolors to our attention. Soon after this, we enlisted the collaboration of David H. Miller, Dean of the School of Humanities and Social Sciences at Cameron University and a recognized authority on Möllhausen and his work. The Museum received a substantial grant from the National Endowment for the Humanities so that Dr. Miller could translate Möllhausen's 600-page Ives expedition diary from German into English for the first time. This important document of American western exploration, with critical commentary by Dr. Miller, is to be published separately. In addition, Dr. Miller has been of inestimable help to this project, providing his knowledge and guidance to Ben Huseman and the other members of the Museum's staff. Without his selfless contribution, this publication, in its present form, would not have been possible.

Even so, the wealth of interpretation and knowledge evident in this volume is largely due to the author himself. Ben Huseman comes to the subject of Möllhausen's travels through his own exhaustive research on the subject. Like his colleague David Miller, Ben's fluency in German and his grasp of the history of German-American cultural interactions make him uniquely qualified to explore his subject, and he has done so here with particular distinction.

Such detailed scholarship often requires careful editing, and once again Nancy Stevens, the Museum's editor, has proved herself worthy of the task. She has worked long hours to make the text flow smoothly for the reader who wants to accompany Balduin Möllhausen on his epic journey. The modern maps that trace the artist's travels are the work of the Museum's talented exhibitions designer, Chris Rauhoff. The color plates and many of the black and white illustrations were produced with great care and professional quality by former and present staff photographers Linda Lorenz, Rynda Lemke, and Steve Watson. Finally, the ease of comprehension for any book is due as much to its design as its readability. As he has so many times in the past, Tom Dawson has taken a wealth of textual and visual material and combined it into a successful whole. To all these people and everyone else who contributed their special skills and talents to make this book a reality, our very special thanks.

RICK STEWART
Acting Director

ACKNOWLEDGMENTS

I could write a book on the making of this book and all the people who have helped since the Museum acquired Möllhausen's watercolors seven years ago, but I will try to refrain from verbosity reminiscent of Möllhausen himself. From the time Harry Stern offered the watercolors to the Amon Carter Museum, Rick Stewart heartily supported my enthusiasm for what I knew to be a great story. I am thankful for his calm confidence, support, and astute guidance, and for the idea of inviting David Miller to assist in the project. Always a good-humored travelling companion, Dr. Miller patiently and kindly shared the vast knowledge, resources, understanding, and contacts that he had built up from years of studying Möllhausen, and he truly measures up to Möllhausen's example in bridging cultural differences. William H. Goetzmann directed my University of Texas master's thesis on Möllhausen and the images of the Whipple expedition, adding to this project much of the intellectual depth that would have been missing without his guidance. My deepest thanks go to him and to Ron Tyler, also of the University of Texas and Director of the Texas State Historical Association, who served as an advisor. Andrew Wallace of the University of Northern Arizona at Flagstaff accompanied David Miller and me as we retraced Möllhausen's route along the Colorado, and at two separate stages he has read the manuscript carefully. I also want to thank Mike Anderson, the owner and skipper of the motorboat that Miller, Wallace, and I used to retrace the routes of Ives' steamboat *Explorer*. Historian Joe Porter of Arlington, Texas, also read the manuscript and offered helpful suggestions.

On the other side of the Atlantic, the late Gerd Möllhausen of Bleicherode, Germany, kindly allowed Dr. Miller and me access to his family's sketches, memorabilia, and other treasures, and we are indebted to his family for allowing us to reproduce many of them

here. Möllhausen biographer Dr. Andreas Graf of Cologne provided much assistance, as did Dr. Peter Bolz of the Berlin Museum of Ethnology; both of these scholars went out of their way to show hospitality and share information. Likewise, Dr. Friedrich Schegk of Munich obtained photographs of the Möllhausen family sketchbooks and kindly provided information. Others who deserve mention here include Dr. Siegfried August of Munich, Dr. Helmut and Inge Krumbach of the Dusseldorf Institute of American Ethnology, Dr. Hans Läng of the Indian Museum of the City of Zurich, Switzerland, and Dr. Axel Schulze-Thulin of the Linden-Museum, Stuttgart.

Descendants and relatives of Baron F. W. von Egloffstein also answered my queries and provided what help they could; these included Winston V. Morrow of Los Angeles and Thomas C. Morrow of Valdosta, Georgia, Dr. Winfrid Freiherr (Baron) von Pölnitz von und zu Egloffstein of Augsburg, and Dr. Albrecht Graf (Count) von und zu Egloffstein of Kunreuth, Bavaria, who even provided a tour of the family's castles.

Many present and former staff members at the Amon Carter Museum were at some point involved in work on the Möllhausen research, exhibition, or publication. Editor Nancy Stevens kept up cheerful encouragement while wading through mounds of manuscript. Former staff photographer Linda Lorenz, a native of Germany, not only made the transparencies of the watercolors but also helped correct my grammar when I wrote letters in German. Also deserving thanks are Jan Muhlert, Rynda Lemke, Steve Watson, Jane Posey, Milan Hughston, Paula Stewart, Sherman Clarke, Tom Southall, and Glynatta Richie.

The staff of the Arizona Historical Society in Tucson were of great assistance to Dr. Miller and me: Director Michael Weber, Steven Horvath, Edward Oetting, Thomas Peterson, Kirsten

Oftedahl, Adelaide Helm, Riva Dean, Lori Davisson, Susan Peters, and Barbara Bush.

Various others around the United States also assisted in the project: William F. Sherman, William Lind, and John A. Dwyer of the National Archives in Washington, D.C.; Kathy L. Dickson of the Oklahoma Historical Society; Richard G. Conn of the Denver Art Museum; Virginia C. Renner and Susan Naulty of the Huntington Library in San Marino, California; Don Batkin, formerly of the Southwest Museum, Los Angeles; Richard Pearce-Moses, Arizona State University Library, Tempe; Carol Brooks, Arizona Historical Society at Yuma; Lisa Martin of the Phoenix Art Museum; Richard Heavily of Northern Arizona University, Flagstaff; Martha Riley, Missouri Botanical Garden Library in St. Louis; Michael R. Green, Texas State Archives, Austin; Barney Lipscomb and Yonie Hudson of the Botanical Research Institute of Texas in Fort Worth; Martin Sarvis of Special Collections, University of North Texas Library, Denton; Curtiss Martin Sr., of the Colorado River Indian Tribes Museum in Parker, Arizona; and Robert W. Karrow and John Aubrey of the Newberry Library in Chicago.

Several graduate students at the University of Texas at Austin offered encouragement and insights—especially Kathleen Dougherty and William Pugsley. A lot of thanks for patience and encouragement goes to my wife Leigh and to our parents.

Among the people who have also contributed to this book in one way or another are numerous librarians and archivists, photographic rights and reproductions specialists, and other specialists too numerous to mention. To all, I extend my heartfelt thanks for their contributions.

BEN W. HUSEMAN

A Prussian Scout of the Wild West

Around the middle of this century, some workmen in Brooklyn were about to raze an old brownstone building when one of them discovered a substantial cache of old books. Among them was an old, worn volume bound in leather and cloth, labeled on the spine as "Original Sketches for the United States Survey of the Territories." After examining the contents, the man decided it might be of some value and took it home. Upon his death in 1982, his relatives sold the volume, which six years later was acquired by the Amon Carter Museum.[1]

Inside the volume were forty-six original watercolors, each measuring approximately 8½ x 11 inches, signed "Möllhausen," and tipped onto a folio-size blank sheet of paperboard, with the title of each painting penciled below. The pictures included a depiction of a steamboat, topographical landscapes along a river, views of deserts, mountains, and canyons, studies of trees, and an intriguing view of a group of natives along a riverbank.

The watercolors were signed by Heinrich Balduin Möllhausen (1825-1905), a Prussian adventurer, traveller, writer, artist, and naturalist who made three lengthy trips to the western territories of the United States in the 1850s. Primarily a writer, he published in German two lengthy travel accounts and over fifty popular novels, short stories, and articles. His first biographer labeled him "the German Cooper" for the similarity between his fiction and American James Fenimore Cooper's, and most of his works involved the American West. To this day, Möllhausen is best remembered in his native land for these writings. For many years his known artworks interested only a small group of ethnologists and historians of the American West. Furthermore, the sum of these works was dramatically reduced during World War II, when over seventy of his watercolors were destroyed in Allied bombing raids on Berlin, Stettin, and Stuttgart. Another group of forty-eight watercolors by Möllhausen remained virtually forgotten in an East German museum and inaccessible to scholars until 1994. Thus the Amon Carter Museum's acquisition of the forty-six long-forgotten watercolors affords a rare opportunity to bring Möllhausen's life and works before the American public.

The subjects of these watercolors clearly relate to Möllhausen's third and last American trip, in 1857-58, when he accompanied Lieutenant Joseph C. Ives' expedition to determine the navigability of the western Colorado River. Ives and his men assembled a specially constructed steamboat near the mouth of the Colorado and slowly travelled upriver almost to the site of present-day Hoover Dam. They then worked their way overland and reached the floor of the Grand Canyon eleven years before the more famous expedition of John Wesley Powell. Möllhausen's watercolors documenting this expedition have considerable historical value as the earliest known views of many sites in the American Southwest, including parts of the Grand Canyon.[2]

For many years, these views were known only through the lithographs and engravings for Lieutenant Ives' *Report upon the Colorado River of the West*, published by the U.S. government in 1861. Some of the prints apparently were derived from the Amon Carter Museum's watercolors, as were a few lithographs in Möllhausen's two-volume diary of this expedition, published in 1861 as *Reisen in die Felsengebirge Nord Amerikas, vom Hoch Plateau von Neu-Mexico, unternommen als Mitglied der im Auftrage der Regierung der Vereinigten Staaten ausgesandten Colorado Expedition* ("Travels in the Rocky Mountains of North America from the high plateau of New Mexico to the Colorado, undertaken as a member of a U.S. government-sponsored expedition," or *Reisen* for short).[3] Although many of the prints after Möllhausen's sketches were stiff, distorted, and dull and many of his surviving works from earlier trips were crude and amateurish, the sketches in the Amon Carter Museum show surprisingly skillful handling of watercolor and gouache; competent drawing, composition, and use of perspective; and fresh, lively renderings of human figures. Thus, the original watercolors provide occasion to reassess Möllhausen's skills as an artist, to study his artistic development and the possible influences he absorbed, and to examine the production of printed illustrations in the mid-nineteenth-century.

Möllhausen did not record dates on a majority of his sketches and paintings, so analyses of subject matter, style, technique, and media are often the only ways of estimating the time of their execution.[4] Since all of his extant artworks relate to his three trips to the United States—each to a different area of the country—subject matter is a useful clue. Although he signed on as a topographic artist with the Ives expedition, Möllhausen approached drawing and painting as a writer or storyteller, and his written accounts and other sources afford useful comparisons with the images. The sketches helped him recall details he might not have written down, and his writings record other details and ideas that he did not or could not sketch, so a full appreciation of his art must consider his background and his written accounts as well as the artworks themselves.[5]

MÖLLHAUSEN'S ARTISTIC DEVELOPMENT 1825-1857

Heinrich Balduin Möllhausen was born in 1825 in Bonn, at that time merely a town in the kingdom of Prussia. His father, Heinrich Möllhausen (b. 1796), was a middle-class building constructor, civil engineer, and former Prussian artillery officer who abandoned his wife and five children when Balduin was about eleven years old. The older Möllhausen went to America, but his son apparently knew few details of his whereabouts.[6] When Balduin's mother, Baroness Elise von Falkenstein, died a year later in 1837, the boy went to live with an unmarried aunt, Adelheid von Falkenstein, and was put under the supervision of a guardian, Count Reinhold von Krassow.

Balduin grew up with an interest in art. According to family tradition, his father had spent a considerable portion of the family's income collecting prints. Balduin received "very good" marks in calligraphy and drawing at the Gymnasium (German secondary school) in Bonn, and by the age of sixteen he had expressed a desire to become a painter.[7] However, his guardian preferred that he not study art, instead sending the young man to Pomerania to study agriculture and become a farmer.[8]

After serving briefly in the Prussian Army in 1845-46 and again during the Revolution of 1848, the young Möllhausen, acting on a long-standing fascination with America, left Europe for the United States in early 1850. His biographers have suggested a number of possible reasons for his wanderlust. His immediate reason may

have been a desire to avoid another tour of duty in the army. He apparently also felt stymied by his guardian's demand that he stick to agriculture or the military, whereas in America he could escape the economic and social limitations imposed by his "orphan" status and pursue his dreams.[9] A longing for his father may have been an inducement as well.

Whatever his reasons for going to America, the artistic and literary works that Möllhausen produced from his travels exhibit many of the prevailing social and intellectual concerns of his time. Despite the personal, narrative, and anecdotal qualities of his works, many display a scientific empiricism that attempts to understand the mysterious unity of nature through its particular elements in the unexplored natural world. Like other Europeans with training in natural history, Möllhausen often approached unfamiliar subjects by attempting to fit them into a Linnaean-style taxonomic system of classification, which he expressed visually in composite images of "types." This European fondness for classification grew out of a wish to understand the fundamental questions of the cosmos—in short, a quest for knowledge. The peoples, animals, plants, and materials of the New World, so unlike anyone or anything encountered in the Old World, were hard to reconcile with European conceptions of static chains of being and hierarchical categorization. These discoveries eventually taught most thinking Europeans that their old beliefs were outdated and unacceptable, but in the meantime, classifying unfamiliar natural phenomena offered hope to Möllhausen and his European contemporaries that they could discover new order in the apparent chaos of New World nature.

Möllhausen's works also exhibit a strong sense of Romanticism, especially the notion of "longing," defined by philosopher Johann Gottlieb Fichte as a "hankering after the unknown" and more recently as "the first and almost last word of German Romanticism."[10] The romantic European quest for the unknown and the exotic had stimulated Möllhausen's imagination. Adventure novels such as Daniel Defoe's *Robinson Crusoe* were among his favorite reading, and works by Washington Irving and James Fenimore Cooper had filled him with Romantic notions about the American wilderness and its inhabitants. He was familiar with the image of the American Indian as the "noble" or "gentle savage," a conception that had permeated German culture through the writings of Rousseau, Chateaubriand, and Herder. He undoubtedly was also familiar with prints of distant lands and exotic peoples, based on the reports and sketches of explorers and travellers. Like other Germans of his time, he imagined America as a land of primeval forests and mighty streams where Europeans, wearied of old traditions and failed revolutions, could begin anew.[11]

In many respects, Möllhausen's romanticism was not a rebellion against the Enlightenment, but an extension of it. Recognizing the importance of the emotions, romantics sought a fusion of feeling and thought rather than the substitution of feeling for thought. Although objective observation was the goal of naturalists like Möllhausen, they realized this was probably impossible because their own perspective was constantly changing. Therefore, they tried to record their own changing visual and mental processes and feelings just as they recorded other objects of nature.

Two of the more interesting Enlightenment ideas reflected in Möllhausen's literary works were toleration, as dictated by reason, and the universal brotherhood of Man. Exploration, by definition, enlarged the horizons of thinking individuals like Möllhausen and impelled them to become world citizens, loosening the grip of nationalistic and local attachments. As he depicted the exotic diversity of peoples and cultures in the American West, Möllhausen also attempted to describe a common humanity and the universal in human nature. His prejudices occasionally prevented him from seeing this, but the fact that he attempted at all differentiates him from many of his less enlightened contemporaries, who looked upon other races as subhuman. Perhaps it is not just coincidence that the German word for Enlightenment, *Aufklärung* (from *aufklären*, "to clarify"), also means "reconnaissance"—the very act Möllhausen engaged in during his three trips to America.

Most of Möllhausen's artworks, including his Ives expedition images, belong in the category of travel and exploration art, part of the great international scientific effort to explore and depict the world's unknown territories before the advent of field photography. These works are pieces of a visual, geographic mosaic of the world assembled by late-eighteenth- and early-nineteenth-century travellers, scientists, and geographers.[12] Like his written works, his drawings and watercolors were informative, didactic, and entertaining. Reflecting his lack of technical training and his capacity for romantic fantasy, his early works sometimes blurred fact and fiction as he responded to the overwhelming and bewildering variety of peoples, animals, plants, and places he encountered, despite his

FIG. 1
Otto Grashof, *Duke Paul von Württemberg,*
oil on canvas, 1853,
photograph courtesy
of Western History
Research Center,
University of Wyoming

scientific intention to portray things as they actually were. As his artistic skills improved with time, however, his works took on a greater illusion of truth and an impetus toward realism, although the exotic realities of the American West always stimulated his imagination.

There are no known extant images that Möllhausen created during his childhood, but a number of works have survived from his first trip to America in 1850-52.[13] He sometimes sketched in graphite or pen and ink with wash, or a combination of these materials, while in the field, and later he composed other works from field sketches or from memory, with the intention of publishing them as prints. Little is known about Möllhausen's exact whereabouts for much of his first trip to America. He arrived in New York in July 1850 and wandered through the Midwest, working as a sign painter, hunter, trapper, and court reporter. Near St. Louis in early 1851, he met Duke Paul of Württemberg (1797-1860), a seasoned world traveller and amateur naturalist, ethnologist, and collector (fig. 1)[14] who hired the young man as an assistant in his travels. They made two lengthy trips together:

FIG. 2
F. Mayer after Duke Paul von Württemberg, *View of the Missouri*,
color lithograph, 1823, The Saint Louis Art Museum, Museum Purchase

one from St. Louis to the Great Lakes via the Mississippi and Illinois Rivers in the summer of 1851, and another beginning in August from St. Louis up the Missouri River and along the Oregon Trail toward the Rocky Mountains.[15] Although Möllhausen's primary duties included hunting, performing camp chores, standing guard, tending animals, and assisting with the collections, the duke—who himself kept sketchbooks—encouraged the young man to draw in what little spare time remained.

Duke Paul's influence in shaping Möllhausen's early artwork cannot be overemphasized. Even some of Möllhausen's later sketches bear some similarity to the duke's topographical style (compare, for example, figs. 2 and 3). The well-educated and experienced duke introduced his assistant to various illustrated travel books,[16] and from their combined perspectives as foreigners, they well understood which subjects were uniquely American and worth recording. The duke intended to reproduce Möllhausen's sketches, along with some of his own, as illustrations for his own account of their travels together.[17] This relationship recalled other aristocratic European adventurers and their artists who had already

FIG. 3
Balduin Möllhausen, *River-Sand-Green bluffs* [*Landscape View of Canadian River*], watercolor and graphite on paper, mounted on paper, 1853, Whipple Collection, Oklahoma Historical Society

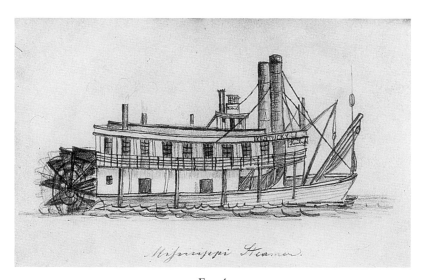

FIG. 4
Balduin Möllhausen, *Mississippi Steamer*, graphite on paper, c. 1850, Sketchbook 1, Möllhausen Family, Bleicherode, Germany, photograph courtesy of Dr. Friedrich Schegk, Munich

FIG. 5
Balduin Möllhausen, *Otoe Nobility*, graphite with sepia ink wash on paper,
1852, Sketchbook 2, Möllhausen Family, Bleicherode, Germany,
photograph courtesy of Dr. Friedrich Schegk, Munich

FIG. 6
Balduin Möllhausen, *Camp in
Wisconsin [Duke Paul or Möllhausen
and Friends]*, graphite on paper,
c. 1851, Sketchbook 1,
Möllhausen Family, Bleicherode,
Germany, photograph courtesy of
Dr. Friedrich Schegk, Munich

travelled the American West—most notably Scottish nobleman
Sir William Drummond Stewart with Baltimore artist Alfred Jacob
Miller (1810-1874) and German Prince Maximilian of Wied-
Neuwied with Swiss artist Karl Bodmer (1809-1893). These earlier
travellers had seen and recorded many important ethnographic
details about Native American life before increased contact with
whites radically changed those ways of life, and they had pictured
many geographical regions for the first time according to western
European traditions of art.[18] Duke Paul undoubtedly had the
examples of Maximilian's famous atlas and Bodmer's illustrations
in mind, and Möllhausen's earliest surviving sketches have similar
subject matter: topographical landscapes, town views, trading posts,
lighthouses, sailing ships, river steamers, Indian portraits, costume
pictures, dance scenes, exterior and interior views of Indian
dwellings, and scenes of hunting and camping (figs. 4-9, 14-19).[19]
Duke Paul and Möllhausen had even covered much of the territory
traversed by these earlier nobleman-artist pairs. The duke may
also have recommended the prints and paintings of artist and
ethnologist George Catlin (1796-1872),[20] who was another important model for Möllhausen, inspiring not only Möllhausen's subject

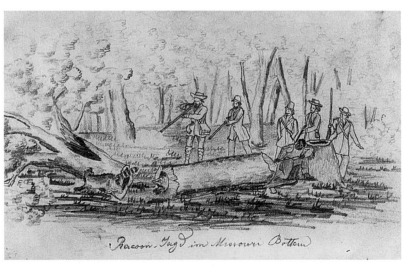

FIG. 7
Balduin Möllhausen, *Raccoon Hunt in the Missouri River Bottom*, graphite on
paper, c. 1851, Sketchbook 1, Möllhausen Family, Bleicherode, Germany,
photograph courtesy of Dr. Friedrich Schegk, Munich

FIG. 8
Balduin Möllhausen, *Negro Ball*, graphite on paper, c. 1851,
Sketchbook 1, Möllhausen Family, Bleicherode, Germany,
photograph courtesy of Dr. Friedrich Schegk, Munich

FIG. 9
Balduin Möllhausen, *Omaha. Katinga-inga*, graphite with
sepia ink wash on paper, c. 1852, Sketchbook 2,
Möllhausen Family, Bleicherode, Germany, photograph courtesy
of Dr. Friedrich Schegk, Munich

matter and compositional types (compare, for example, figs. 9 and 10) but also his sense of urgency about recording the vanishing frontier and traditional Indian ways of life and his criticism of common Anglo-American attitudes toward many tribes.[21]

Möllhausen's early sketches are crude, naive, and amateurish, yet some of his genre scenes of American life possess charm, wit, humor, "coziness," and a *joie de vivre* (or *Lebensfreudigkeit*) typical of German middle-class art of the period.[22] His perspective is generally bad, with skewed angles, distorted ellipses, and sometimes a poor sense of scale. The human figures usually lack proper proportions and are a bit stiff, with facial features that are often caricaturish or overly simplified. (Curiously, these same adjectives have been used to describe some of the fictional characters in his novels as well.) While the disproportionately tiny feet and hands of all of his figures (especially see figs. 5, 8, and 9) derive in part from the prevailing ideal human type in European fashions of the 1840s and 1850s—which tended to de-emphasize these parts of human anatomy (see fig. 11)—some of the problem obviously lies with the young artist's failure to observe nature closely.

Certain sketches pertaining to Möllhausen's first American trip—particularly the ones executed in pen and ink, with their emphasis on line (fig. 12)—derive from the neoclassical "outline style" popular in European and American book illustration during the first half of the nineteenth century. Möllhausen employed this style, which originated in late eighteenth-century England in the work of artist John Flaxman[23] and was inspired by antique Greek vase painting, in his drawings of Indians, as had Catlin and the American illustrator F. O. C. Darley (fig. 13). They may have thought that the style associated with the classical myths of the European past was ideal for rendering the American "noble savage"—a way of associating a European's own origins with the "primitive" Native American.

FIG. 10
George Catlin,
engraving from
*Catlin's . . . North American
Indian Collection*, 2 vols.
(London: G. Catlin, 1848),
Amon Carter Museum Library

FIG. 12
Balduin Möllhausen, *Oglala Sioux Hunting Buffalo*, ink on paper, c. 1852,
Peabody Museum, Harvard University

FIG. 11
After Grandville, *La Revanche, ou les français au Missouri*, lithograph, 1844,
Amon Carter Museum, Fort Worth. Native Americans, who had been
"on display" in Europe, had their opportunity for revenge by pointing
out the absurdities of European fashion.

FIG. 13
After Felix O. C. Darley, untitled lithograph from
Scenes in Indian Life, 1843, Amon Carter Museum

Fig. 14
Balduin Möllhausen, *Dacotah [Möllhausen Demanding the Return of His Knife]*, graphite on paper, c. 1851-52, Sketchbook 2, Möllhausen Family, Bleicherode, Germany, photograph courtesy of Dr. Friedrich Schegk, Munich

Fig. 16
Balduin Möllhausen, *Lodge of the Otoe Indians*, graphite with sepia ink wash on paper, 1852, Sketchbook 2, Möllhausen Family, Bleicherode, Germany, photograph courtesy of Dr. Friedrich Schegk, Munich. Möllhausen is reclining at right, inside the lodge.

The experiences with Duke Paul had a profound impact upon Möllhausen's life and subsequent career, and many sketches of their adventures together demonstrate the young artist's flair for dramatic narrative. Hair-breadth escapes became an almost regular occurrence for the two Germans on their trip together back from Fort Laramie in what is now Wyoming. One of Möllhausen's sketches, typical of the "encounter" genre in exploration and travel art, shows the artist surrounded by a party of Sioux on October 23, 1851 (fig. 14), and another depicts both

Fig. 15
Balduin Möllhausen, *Cayouas [Kiowas] Indians*, graphite on paper, c. 1851-52, Sketchbook 2, Möllhausen Family, Bleicherode, Germany, photograph courtesy of Dr. Friedrich Schegk, Munich

men surrounded, this time by Kiowas on October 26 (fig. 15).[24] One of the best-known and most harrowing experiences in Möllhausen's eventful biography occurred several days after these incidents, when he and the duke were separated during severe early winter snows. The duke fortunately caught a ride with a heavily loaded mail coach, but Möllhausen spent five more weeks on the plains of Nebraska, alone, half-starved, nearly frozen, and under threat of hostile Indians. Rescued in early January 1852 by a passing band of friendly Otoes, he recovered for several weeks in their village near Council Bluffs, Iowa, and attempted to learn about their way of life (figs. 16-17).[25] Needless to say, these experiences both fulfilled and forever changed many of his romantic expectations about "noble savages."

Möllhausen drew many sketches of these adventures while recuperating for several more weeks in February 1852 in Bellevue, Nebraska, at the American Fur Company's popular trading post (fig. 18) run by Pierre Sarpy.[26] (Later, in New Orleans, Möllhausen

met Duke Paul and there evidently made copies of some of the sketches for his patron.[27]) He obviously intended certain sketches to be factual representations. For example, his view of Fort Robidoux is the only extant image of that private trading post (fig. 19), whose site historians located in Scott's Bluff, Nebraska, using a photograph of the sketch.[28] Sketches of Möllhausen's Oto rescuers, their dwellings, and the trading post at Bellevue probably also belong in this category. Other scenes could have existed only in the artist's imagination but may have been inspired by his personal experiences; for example, a sketch titled *Pocahontas Saving Capt. Smith's Life* may relate indirectly to his unrequited romantic attraction for Emily Papin, a beautiful half-blood French-Pawnee girl he met at Bellevue. He may have relied on prints after earlier western artists as well; his sketch of a "scalp dance of the Otoes" is similar to a print of a scalp dance after Catlin (figs. 20 and 21), and an image of Indians hunting buffalo resembles one by Swiss-born Peter Rindisbacher (figs. 22 and 23). Prints after Rindisbacher's painting

FIG. 17
Balduin Möllhausen, *Otoe Squaw [with Horse Travois]*, graphite with sepia ink wash on paper, 1852, Sketchbook 2, Möllhausen Family, Bleicherode, Germany, photograph courtesy of Dr. Friedrich Schegk, Munich

FIG. 18
Balduin Möllhausen, *Sarpy's Establishment, Bellevue*, graphite with sepia ink wash on paper, 1852, Sketchbook 2, Möllhausen Family, Bleicherode, Germany, photograph courtesy of Dr. Friedrich Schegk, Munich

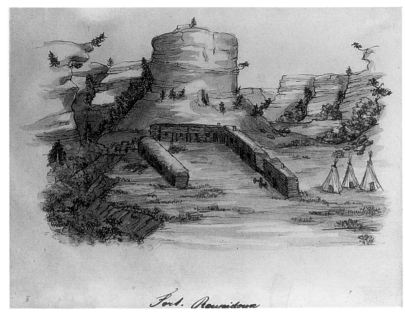

FIG. 20
Balduin Möllhausen, *Scalp Dance of the Otoes*,
c. 1852, Sketchbook 2, Möllhausen Family,
Bleicherode, Germany, photograph courtesy
of Dr. Friedrich Schegk, Munich

FIG. 19
Balduin Möllhausen, *Fort Robidoux*, ink and wash on paper, c. 1851,
Sketchbook 2, Möllhausen Family, Bleicherode, Germany, photograph
courtesy of Dr. Friedrich Schegk, Munich

were quite common, and Möllhausen noted that even at Sarpy's trading post at Bellevue, "several old lithographs of Indian portraits decorated the unfinished log walls."[29] Möllhausen probably never witnessed a scalp dance or a buffalo hunt firsthand, but he does claim in his journal to have briefly participated in a "horse dance." Interestingly, although he reported in his journal that he removed himself when the celebrants began drinking whiskey and shooting off their firearms, for fear a shot might go astray, his sketches portray the Otoes as he fancied them rather than as drunken revellers. Möllhausen's inclination to classify Native Americans by "types" in his sketches (see figs. 5 and 9 for early examples) led to stereotyping, but it also reflected the importance of tribal identity on the frontier, where mistakes could prove fatal. Ethnic "types" would appear over and over in his work, but when Möllhausen knew the names of his sitters, he wrote them down, suggesting that he could sometimes move beyond simple stereotypes to regard Indians as individuals.

FIG. 21
After George Catlin, *Scalp Dance*, from *Catlin's Indians: The Kemper Portfolio in the Collection of the Albrecht Art Museum* (St. Joseph, Mo.: The Museum, 1990), courtesy of Stanford University Library

FIG. 22
Balduin Möllhausen, *Oglala* [*Sioux Hunting Buffalo*],
graphite with ink wash on paper, c. 1851-52, from Sketchbook 2,
Möllhausen Family, Bleicherode, Germany, photograph courtesy
of Dr. Friedrich Schegk, Munich

FIG. 23
After Peter Rindisbacher, *Hunting the Buffalo*, lithograph, 1833,
Amon Carter Museum

FIG. 24
Eduard Hildebrandt, *Alexander von Humboldt in His Study*,
lithograph, n.d., Corcoran Gallery, Washington, D.C.

In the fall of 1852, the Prussian consul in St. Louis hired Möllhausen to accompany a shipment of wild animals destined for the Berlin Zoo. Arriving safely home the next January, he soon came to the attention of several important people in Prussian scientific, cultural, and government circles, including the director of the Berlin Zoo, Dr. Hinrich Martin Lichtenstein (1780-1857). A naturalist who had previously worked as a professor of medicine and zoology, Lichtenstein was quite active in scientific and cultural circles in Berlin and had published, in addition to various other scientific works, a two-volume illustrated book on his travels in southern Africa, titled *Reisen im südlichen Afrika* (1810, 1811). Quite impressed with Möllhausen, Lichtenstein referred him to Alexander von Humboldt (1769-1859), the famous explorer, geographer, and scientist (fig. 24) whose extensive publications contained many prints based upon his own sketches.[30] The eighty-four-year-old Humboldt was a serious proponent of visualization to organize collected facts: his great atlas of Latin America (1811)

Fig. 26
After Alexander von Humboldt,
Costumes des Indiens de Mechoacan,
aquatint and etching, from
Alexander von Humboldt,
*Vue des Cordillères, et monumens des
peuples indigènes de l'Amerique*
(Paris, 1810), courtesy of
Special Collections, University
of North Texas Library

Fig. 25
After Alexander von Humboldt, engraving and etching from *Vue des
Cordillères, et monumens des peuples indigènes de l'Amerique* (Paris, 1810),
courtesy of Special Collections, University of North Texas Library

included maps, topographical landscapes, geological cross-section diagrams, and ethnographic studies of particular taxonomic "types" of people and costumes (figs. 25 and 26); other volumes contained detailed botanical studies. These kinds of illustrations and descriptions comprised a new discipline, which he had termed "physical geography," and whose ultimate goal was a large-scale survey of nature.[31] As the premier universal scientist-traveller,[32] Humboldt had already influenced Prince Maximilian of Wied-Neuwied, Duke Paul of Württemberg, and others whose works Möllhausen had studied, and he soon became Möllhausen's most important role model, mentor, and patron.

Together, Lichtenstein and Humboldt introduced Möllhausen to the members of the Royal Geographical Society in Berlin, enlisting subscribers to help send the young man back to the United States as a travelling collector in natural history. They provided him with contacts, letters of recommendation, training, and access

to important books and Berlin collections. In Prussia, Humboldt (and probably also Lichtenstein) could directly advise Möllhausen on almost every aspect of his possible future duties, from reading scientific instruments to collaborating with printmakers and publishers. Möllhausen also spent hours in Humboldt's personal library. As Lichtenstein wrote to the Prussian Minister of Culture, Karl Otto von Raumer, in a few months the young man had "familiarized himself in all the royal natural history collections with the methods of collecting, preparing, and especially also with the instructions for the excavation and difficult preservation of fossilized skeletons" and "also perfected his drawing."[33] At this time Möllhausen fell in love with and became engaged to Humboldt's private secretary's daughter, Caroline Seifert, who lived in Humboldt's household; her father, Johann Gustav Seifert, who had once accompanied Humboldt to Siberia, oversaw virtually everything in Humboldt's Berlin household. The high probability that Caroline was actually Humboldt's illegitimate daughter[34] probably also influenced Humboldt's continuing support and personal interest in the young artist. From his position as Chamberlain, cultural advisor, and Privy Councilor, Humboldt also got

Möllhausen an audience in April with King Friedrich Wilhelm IV (who ruled Prussia from 1840-58).[35]

The careers of men like Möllhausen and even Humboldt owed much to the king.[36] Today remembered primarily for his conservative politics and vacillating role in the Revolution of 1848, the king supported and encouraged the arts and sciences, himself painted in watercolors and collected art, and, at Humboldt's urging, sponsored German artists on lengthy expeditions to the New World that built up quite a collection of exploration and travel art. Although the king and Humboldt especially favored sending artists to Central and South America, any locations whose visual aspects were little known might qualify for state-assisted endeavors.[37] One problem, however, was finding artists with adequate training who would undertake the risks and could endure the hardships of such travel.

Möllhausen was willing to return to America, and in mid-April 1853, even before learning that the king would employ him in the Prussian government, he was off on his second trip to the United States, entrusted with dispatches from the American embassy at Berlin and provided with letters of recommendation and financial backing from his high-placed friends. Humboldt wrote of him:

> Instructed by the intimate relations which he enjoys with the learned naturalists of my country, acquainted with the deficiencies of the Museums, by his intelligent zeal and his courageous and enduring activity, he will be very useful as a Collector for the natural history of savage animals and the study of minerals and rocks. A very remarkable talent for drawing as applied to picturesque sites and scenes of Indian life, will add to the fruits of his distant voyages.[38]

Such a recommendation from Humboldt carried much weight around the world and especially in the United States, where in 1804 Humboldt had visited with such scientific luminaries as Benjamin Smith Barton, Dr. Benjamin Rush, Charles Willson Peale, and President Thomas Jefferson, and where many of his publications were still well known and widely read. However, in an age of increasing specialization, Humboldt was himself already becoming something of an anachronism, as scientific knowledge, growing at an ever-multiplying rate, became impossible for one man—even one of Humboldt's education and intellect—to master.

Möllhausen was hardly a Humboldt. By the time he left Prussia for the United States on his second trip, he had absorbed only enough knowledge to become a sort of dilettante in science and art. By chance, however, when he arrived in Washington, D.C., in May of 1853, there were considerable opportunities for artists to do fieldwork in the nation's western territories. Congress was ready to send scientific expeditions to explore four principal routes for a transcontinental railroad from the Mississippi River to the Pacific Ocean, and each expedition would require the services of an artist and other specialists.[39] Throughout much of its hitherto short history, the U.S. government had sponsored scientific exploration in its vast western territories and had hired civilian artists and scientists to accompany these expeditions. No artist had been with Lewis and Clark in 1804-06, but two—Samuel Seymour and Titian Peale—had accompanied Major Stephen Long's exploring expedition to the Rocky Mountains in 1819-20.[40] Since that time, and especially in the 1840s, official U.S. expedition reports often contained prints depicting landforms, flora, fauna, fossils, items of archaeological interest, natives, and settlements or villages, all based on sketches by expedition artists. Since all the data gathered on these expeditions was thought to be somehow related—part of a great cosmic system—the government survey parties recorded all sorts of trivia and brought back thousands of rocks, plants, animals, and ethnological artifacts, leaving it to the artist to record whatever could not be collected or shipped back east.[41] Sometimes artists could decide what they wished to sketch (scenes that could be described as "picturesque" or "sublime" often qualified as suitable subjects), but occasionally the commander of the expedition or one of the scientific specialists would direct the artist.

With recommendations from Humboldt and Leo von Gerolt, the Prussian ambassador in Washington, Möllhausen found a position as "topographer and draughtsman" with the expedition commanded by Lieutenant Amiel W. Whipple (1817-1863) of the Army's Corps of Topographical Engineers, who was to explore a route west along the thirty-fifth parallel.[42] Few documents specifically define Möllhausen's tasks on the expedition, for which he was to receive a salary of $1200 per annum.[43] According to Whipple's instructions, he was to keep "an official journal," making "such drawings and memoranda" as he deemed to be of value or interest for the expedition.[44] This apparently allowed Möllhausen considerable personal artistic discretion, and before leaving Washington, he made drawings of animals from their skins for scientists at the Smithsonian Institution, who also instructed him regarding the

FIG. 27
Balduin Möllhausen, *Camp Wilson and Fort Smith*, graphite on paper, 1853,
Whipple Collection, Oklahoma Historical Society, photo 9256-OD14

FIG. 28
Balduin Möllhausen, *Mountain Peaks [La Cuesta Sept. 28, 1853]*,
graphite and gouache on paper, 1853, Whipple Collection,
Oklahoma Historical Society, photo 9256-P59

natural history collections he was to make in the field.[45] Although the job of topographer and draftsman required some of the kind of artwork that Möllhausen had done for Duke Paul, the artist was freed of cooking and other menial chores he had performed for the duke; as a senior member of the expedition, he had his own personal tent and access to a shared cook and several servants.[46] Möllhausen still had to devote much of his time to natural history studies and specimen collecting, but he referred to himself as "the German naturalist," perhaps to indicate that he did not primarily perceive himself as a U.S. government artist, and also sketched anecdotal incidents beyond the scope of his official duties.

In mid-June 1853, Möllhausen arrived in Fort Napoleon, at the mouth of the Arkansas River. Travelling with Whipple from Fort Smith through present-day Oklahoma, Texas, New Mexico, Arizona, and California, he sketched landscapes, landmarks, settlements, camp scenes, a river crossing, utensils, artifacts, pictographs, scenes of daily life, and renderings of various botanical and zoological specimens (figs. 27-32 and cats. 18d-e). He depicted Indian subjects from a number of tribes, including the Choctaw, Shawnee, Delaware, Comanche, Waco, Kiowa, Pueblo, Navaho, Apache or Yavapai, Chemehuevi, Mohave, and Cahuilla—and would record

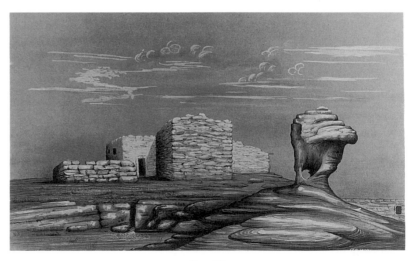

FIG. 29
Balduin Möllhausen, *Covero*, ink wash with opaque white and graphite on brown paper, 1853, Whipple Collection, Oklahoma Historical Society, photo 9256-ODU8

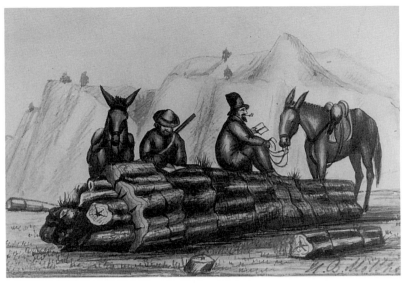

FIG. 30
Balduin Möllhausen, *Petrified Tree near Lighodendron Creek*, pen and
ink wash with graphite on paper, mounted on paper, 1853,
Whipple Collection, Oklahoma Historical Society, photo 9256-ODU12

many of them again on the Ives expedition (see figs. 53-54 and p. 132).[47]

Möllhausen drew most of his sketches in the field, or during a stopover in Albuquerque in the middle of the expedition (from October 5-November 8, 1853), or during his stay in Washington, D.C., between early May and August 1854, after the expedition ended.[48] Though quite similar in execution to his better work from his earlier trip, these sketches reveal a steady development of his artistic skills. Working mostly in graphite, on sheets of paper about 9 x 14½ inches in size, he employed a greater variety of techniques, using lighter and heavier pencil strokes, different grades of graphite, and sharp and dull points to create emphasis and contrast between details; occasionally he rubbed the graphite with either his hand or a smudge stick to achieve tonal effects, applied a monochromatic ink wash, or added watercolor and gouache.[49] The general absence of color in most of his sketches would have aided the American copyists, who translated them into monochromatic woodcuts or lithographs with only a few colors for the official reports of the expedition.[50]

The sketches from Möllhausen's second trip to America also reveal his continuing artistic weaknesses. His modeling was awkward when he worked with watercolor and gouache, a medium in

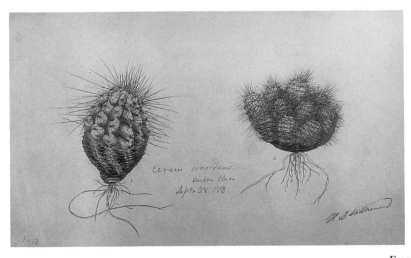

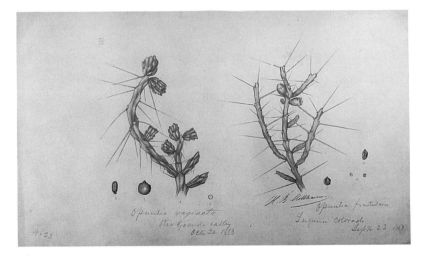

FIGS. 31-32
Balduin Möllhausen, *Cereus condeus* (l.) and *Opuntia vaginata and Opuntia fruteseus* (r.), graphite on paper, 1853, Whipple Collection,
Oklahoma Historical Society, photos 9256-BSD13 and 9256-BSD23

which he apparently had little prior experience, and he generally created highlights by using opaque white on toned paper rather than leaving areas of white paper blank (see fig. 28 and cat. 19d). Although his figure drawing had improved, with poses that seemed more relaxed and natural, he still had difficulty with proportions, and some of his Indian faces look identical or lack individualized features. He was also still plagued by a lack of knowledge about the rules of perspective and by a poor sense of scale.

Prior to the expedition and immediately afterwards, Möllhausen reviewed the best available illustrated books on the western territories and their inhabitants, with illustrations by George Catlin, Seth Eastman, John Russell Bartlett, Henry C. Pratt, and others.[51] Previous U.S. government expeditionary reports were another important source of inspiration, many illustrating subjects similar to those he encountered along Whipple's railroad survey route. Möllhausen cited these works in his published *Diary*, and during the expedition itself, he could have referred to some that were among Lieutenant Whipple's baggage.[52] Although he seldom copied their compositions exactly, some of Möllhausen's works strongly resemble these earlier prints. The lithographs after artist John Mix Stanley (1814-1872), in Lieutenant William H. Emory's report of a Mexican War reconnaissance along much of Whipple's route, influenced Möllhausen's drawings,[53] as did Lieutenant James W. Abert's illustrations of reconnaissances through Comanche country (published in 1846) and New Mexico (published in 1848).[54] Expeditionary artist Richard Kern (1821-1853), whom Möllhausen could have met in Washington, D.C., prior to the expedition, was another likely influence[55]; both Whipple and Möllhausen specifically mention Captain Lorenzo Sitgreaves' *Report of an Expedition down the Zuñi and Colorado Rivers* (1853), with prints after Kern.[56] Kern had anticipated Möllhausen in his sketches of the pueblo of Zuñi, San Francisco Mountain and other geological formations, springs, petrified stumps of trees, inscriptions on Inscription Rock, pottery fragments, and members of various Southwestern tribes such as the Navaho and the Mohave (fig. 33). Kern's death in April 1853 at the hands of Paiute Indians in the Utah Territory also dramatically underscored the hazards of an expeditionary artist's work (although Möllhausen did not learn about this event until late in the Whipple expedition).

In Washington, D.C., sometime during his second American visit, Möllhausen apparently met John James Young, a copy artist

Fig. 33
After Richard Kern, *Mohave Indians (Big Colorado N.M.)*, lithograph, from Lorenzo Sitgreaves, *Report of an Expedition down the Zuni and Colorado Rivers*, Senate Exec. Doc. 59, 32d Congress, 2d Session (Washington, 1853), Amon Carter Museum

who would determine much about how the American public perceived Möllhausen's artwork. Young (c. 1830-1879) was a Prussian-born civilian draftsman employed by the Office of Explorations & Surveys to copy, rework, and complete sketches by the expedition artists so they could be reproduced as illustrations for published official government reports.[57] The complicated process of producing finished prints from the expeditionary artists' sketches took considerable time, and Möllhausen's stay in Washington in the summer of 1854 was too brief to allow him to do his own work, so Young evidently translated some of Möllhausen's sketches for the printer.[58]

None of Young's copy watercolors of Möllhausen's sketches for the Whipple expedition survive, but inscriptions on several lithographs in Whipple's report credit Young with redrawing works by Möllhausen, by Albert H. Campbell, the expedition's assistant engineer and surveyor, and by Lieutenant John C. Tidball, a military escort commander.[59] Young's copy watercolors of Tidball's field sketches (figs. 34 and 35) reveal that he did not copy the originals exactly, but instead redrew and rearranged figures and other *staffage* (peripheral matter) in order to create better compositions.

Fig. 34
John C. Tidball, *Last Gate, Bill Williams Fork*, graphite on paper, mounted
on paper, 1854, Whipple Collection, Oklahoma Historical Society,
photo 9256-ODU16

Fig. 35
John James Young after John C. Tidball, *Canon of Bill Williams Fork*,
watercolor and ink on paper, c. 1855, Whipple Collection,
Oklahoma Historical Society, photo 9256-ODU14

Although Young often seemed more skillful than the expeditionary artists in handling the medium of watercolor, he repeatedly lost or altered certain details in the process—as he would when he redrew many of Möllhausen's sketches from the Ives expedition.

In his apparent haste to return to Prussia, Möllhausen may have abdicated part of his responsibilities as a survey artist, since he was unavailable for close collaboration with the other Whipple expedition members and was unable to approve or disapprove of the proof prints produced for the official reports. After he turned in his sketches, he evidently had little, if any, direct involvement in the process, through Lieutenant Whipple, the War Department's Office of Explorations and Surveys, the Smithsonian Institution, or the numerous lithographic and engraving establishments who contracted to produce the illustrations.[60] Unfortunately, careful attention to this post-expedition work was vital not only to the real or perceived success of the expedition itself but also to Möllhausen's reputation as an artist, because the prints made from his original pictures were the only images most of the public saw.

Back in Prussia from early 1855 until 1857, Möllhausen entered a very productive period. He married Caroline Seifert in February 1855, and by the end of that year their first son, Alexander, arrived. He continued to enjoy the friendship and benevolent interest of Humboldt and Lichtenstein, who again gave him access to their personal libraries and introduced him to artists and scholars with similar interests, including George Catlin, who stopped in Berlin in September 1855.[61] Four weeks before his marriage, Möllhausen also began work at his new post as "Custodian of the Royal Libraries in and around Potsdam"—a virtual sinecure granted by King Friedrich Wilhelm IV. This gave him access to important people and to books with rare and expensive illustrations, as well as time to pursue his personal artistic interests. Möllhausen wrote to his American friend, Dr. Spencer Baird of the Smithsonian, that his new job in Potsdam "is not of so fatiguing a nature, I have time enough to draw to paint to learn and to study, and I don't allow a minute to pass unnoticed."[62]

FIG. 36
After Möllhausen, *Navajos*, toned lithograph, from *Reports of
Explorations and Surveys* (Washington, 1856), vol. 3, part 3, opp. p. 31,
Amon Carter Museum Library

Möllhausen continued to produce work relating to his two
earlier American trips. Painting finished watercolors from the
sketches he had brought from the United States, he apparently
hoped to convey more effectively what he had seen and experienced in America to his family, friends, sponsors, and other
interested individuals in Prussia. These watercolors served as
better illustrations than his graphite sketches when he made
presentations before the Berlin Geographical Society and the King
of Prussia, and they were also good accompaniments to his personal diary of the Whipple expedition, which he reworked by late
November 1855 into a manuscript suitable for publication.

As he reworked his sketches, Möllhausen earnestly sought to
improve his artistic skills through practice, study, and no doubt by
viewing paintings and drawings in the various public and private
Prussian collections. His experience with the Whipple expedition
had convinced him of his own artistic inadequacies—perhaps in
part from negative remarks about his artwork and artistic performance on the expedition. Although Lieutenant Whipple characterized Möllhausen's sketches of cacti as "capital drawings,"[63] the
official published report on the Indians of the expedition (1856),

which Whipple co-authored with Thomas Ewbank and Professor
William W. Turner, implies several failings in the artist's renderings
of Native Americans. For example, the authors commented that a
lithograph after Möllhausen's sketch of two mounted Navahos
(fig. 36) "is given as furnished by the artist; though, excepting the
striped blanket of Navajo manufacture, the portraits differ little
from those of the Pueblo Indians. One is represented with hair
cut squarely in front to the eyebrows—a custom not heretofore
attributed to any of the Apache race."[64] Similarly, an illustration
of two "Huéco Indians" lacked the "high cheek-bones, and a wild
look, (which the artist has failed to represent,) totally different
from the quiet features of those representing the preceding
tribes."[65] Although these criticisms could apply to the printmaker
rather than to the original artist, Möllhausen was embarrassed by
the prints that appeared in Whipple's published report. When he
returned to the United States in 1857, he brought along samples
of his more recent works and commented: "I don't like it much to
see my name under the horrible lithographs in Capt. Whipple's
report."[66]

While in Prussia from 1855 to 1857, Möllhausen had a greater
opportunity to absorb Humboldt's ideas on art, which derived from
physical geography and a belief that art should be a creative but
accurate portrayal of nature. *Cosmos* (1845-62), Humboldt's bold
attempt to describe the physical universe and man's knowledge
of it, included critical summaries of the histories of science,
exploration, travel description, and landscape painting, and
emphasized the interrelationship of these endeavors, particularly
as each excited the imagination, expanded knowledge, and enlarged human fields of view. As early as 1808, in *Aspects of Nature*,
Humboldt had called for European artists to visit distant lands,
sketching on the spot and creating accurate, artistic representations of exotic plants and trees to help scientists and geographers
understand the vegetation in various climate zones. Thus, according to Humboldt, artists depicting mountains should pay attention
to borders of growth, snow lines, the height of peaks, geological
structures, outlines of mountain silhouettes, and the colors of
rocks, because such accurate images would permit comparative
studies of mountains and provide insight into their geological
make-up.[67]

Möllhausen need not have learned all of these ideas directly
from Humboldt, because their echoes could be found elsewhere

in German art. For example, Carl Gustav Carus (1789-1869), a follower of the great German romantic master Caspar David Friedrich, was a physician and naturalist as well as a philosopher and painter. Carus viewed the goal of landscape painting as the visualization of the inner workings of geological phenomena, or *Erdlebenbildkunst* (pictorial art of the life of the earth), so the viewer could determine the physical properties of a particular rock in a picture and deduce its geological history (fig. 37).[68] In his influential *Nine Letters on Landscape Painting* (1831 and 1835), he also articulated the romantic's understanding of nature:

> Man, when he contemplates the magnificent unity of a natural landscape, is made conscious of his own insignificance and, feeling that everything is a part of God, he loses himself in the Infinite and renounces his individual existence. To bury oneself thus is no loss, it is a gain; what one can ordinarily see only through the spirit almost becomes visible to the naked eye; one is convinced of the infinite universe.[69]

FIG. 38
Eduard Hildebrandt, *Rio de Janeiro, April 1844*, watercolor, Staatliche Museen zu Berlin–Preussischer Kulturbesitz Kupferstichkabinett

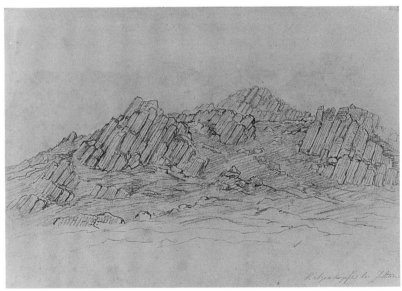

FIG. 37
Carl Gustav Carus, *Die Katzenköpfe bei Zittau*, graphite on paper, 1820, Staatliche Kunstsammlungen, Kupferstich-Kabinett, Dresden, Germany

Another influential artist was Eduard Hildebrandt (1818-1869); also a favorite of Humboldt and Friedrich Wilhelm IV, he gave Möllhausen some artistic training. Hildebrandt was a world traveller whose field sketches and finished works met Humboldt's desire for the "true" reproduction of physical nature (fig. 38) and a master watercolorist whose technical virtuosity with transparent washes and brushstrokes suggested sureness of touch and the ability to work at great speed.[70] He knew how to include subtle details of scientific and picturesque interest in his compositions without overburdening them, and with his gifted sense of color, he could convey the light and the atmospheric effects of tropical or wintry climates. He also had an excellent sense of perspective, human anatomy, and proportion. In short, Hildebrandt was well suited to train Möllhausen for the kind of total geographical reportage expected of him, and his influence evidently coincided with some major changes in Möllhausen's artworks in late 1854 or 1855.

The most pronounced of these changes was Möllhausen's increasing use of the medium of watercolor and gouache, which helps to date a number of works whose subjects relate to his first trip to the United States but whose media and style are like his later works.[71] In general, the finished pictures that Möllhausen produced after returning to Prussia are much more refined than the ones he produced in the United States. The watercolor and gouache works he created in America for the Whipple report have a preponderance of opaque paint, while comparable works painted in Germany use more transparent washes and a more skillful, selective application of paint. Further, Möllhausen painted many of the later works on a medium-rough watercolor paper, again suggesting greater familiarity with the medium, the influence of Hildebrandt's instruction, and possibly greater availability of art supplies and increased ease of painting in the more comfortable surroundings of the Fatherland.

Comparing these watercolors with his earlier sketches demonstrates Möllhausen's continuing progress in technique, composition, and drawing skills. His watercolor of *Fort Robidoux*, probably dating from 1855-57, demonstrates some rudiments of perspective that were missing from the earlier ink and wash drawing, c. 1851, in his second sketchbook (compare fig. 39 with fig. 19). In the sketch he depicted the fort's log blockhouses from above, unintentionally making them appear to tilt into the foreground, while in the watercolor the same structures are placed in a more convincing perspective, creating a better sense of scale, distance, and integration with the surrounding landscape. He also used the comparatively rough texture of the watercolor paper to give tactility to the rocky land formations and foreground soil, rocks, and grasses—an effect entirely missing in the earlier sketch.

Another example of Möllhausen's improved perspective is his watercolor composition, c. 1855-57, of Sioux Indians by a campfire (fig. 40), with large foreground figures and smaller horses suggesting linear perspective and the dark foreground and lightly painted horizon providing a sense of aerial perspective. This is a marked improvement over the earlier sketch titled *Dacotah [Möllhausen Demanding the Return of His Knife]* (fig. 14), in which warriors in the foreground are somewhat smaller than those in the middle ground, and the equivalent tone of the foreground and background flattens the composition. The gestures of the figures of the later work also seem less stiff.

At short intervals in the mid-1850s, Möllhausen sold and delivered three portfolios of sketches from his first two trips to

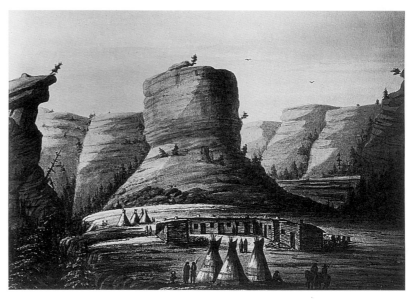

FIG. 39

Balduin Möllhausen, *Fort Robidoux*, watercolor, c. 1855-57, destroyed 1945, formerly in the Staatliches Museum für Völkerkunde, Berlin

America—a total of forty-nine watercolors—to the King of Prussia. These works, long thought lost, recently have been located in the Royal House Library Collection of Her Majesty Queen Elisabeth in the Neues Palais, Potsdam. These watercolors, like the ones in the Berlin Museum of Ethnology, contain gouache in addition to watercolor, and their coloration is generally subdued. Some of these works are identical or similar to known works in other collections, but others are completely unique examples.[72]

Humboldt was able to report on November 7, 1855, that Möllhausen's artwork continued to receive personal attention from the king:

> the talented young Möllhausen who had experienced so many great adventures (from Missouri to San Francisco in 16 months by horse) . . . has properly developed his artistic talent for drawing (landscapes, animals, forests, deserts) through Hildebrandt's help. He just comes from the King, who had summoned him this morning, in order to view his collection of new drawings for forty-five minutes and to overwhelm the

FIG. 40
Balduin Möllhausen, *Sioux Indians by a Campfire*, watercolor,
c. 1855-57, destroyed 1945, formerly in the Staatliches Museum
für Volkerkunde, Berlin

former corporal with well wishes . . . and friendly praise. Rarely is a job so successful.[73]

In his fourth audience with the king, on November 23, 1856, Möllhausen displayed his manuscript diary of the Whipple expedition, along with watercolors and three maps. He probably had completed all or most of the original artwork for his published diary, *Tagebuch*, by February 22, 1857, when Humboldt sent the collection of sketches and most of Möllhausen's manuscript to his friend and government advisor Alexander Mendelssohn, the father of an important publisher.[74] The award-winning Berlin firms of Storch & Kramer and Winckelmann & Sons produced the lithography for the German edition, while Hanhart of London lithographed the illustrations for the English edition (figs. 41 and 42).[75] A Dutch translation was also prepared, and all three editions appeared in 1858, while Möllhausen was on his last trip to the United States.[76]

As a rule, the European-produced prints for Möllhausen's Whipple diary were far superior to the American-produced prints

in the official Whipple report. The embarrassing appearance of the *Report*'s prints probably convinced the artist and his patrons that the diary required the finest quality reproductions. Many of the prints for Möllhausen's book were actually chromolithographs employing four or more separately colored stones, but Möllhausen also furnished the European lithographers with better, more complete original artwork. Since this was a more personal project, he or Humboldt also may have paid better attention to and personally exerted some control over proof prints.

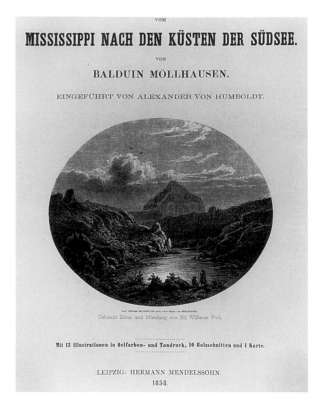

FIG. 41
Eduard Hildebrandt after Balduin Möllhausen, *Colorado River und Mündung von Bill Williams Fork*, wood engraving, from Möllhausen, *Tagebuch vom Mississippi nach den Küsten der Südsee* (Leipzig: Mendelssohn, 1858), title page

FIG. 42
After Balduin Möllhausen, *Camp of the Kioway Indians*, chromolithograph, from Balduin Möllhausen, *Diary of a Journey from the Mississippi to the Coasts of the Pacific with a United States Government Expedition*, 2 vols. (London, 1858), vol. 1, opp. p. 212, Amon Carter Museum Library

FIG. 43
Balduin Möllhausen, *Kiowa Camp*, watercolor and gouache on paper, after 1855, Möllhausen Family, Bleicherode, Germany, photograph courtesy of Dr. Friedrich Schegk, Munich

THE IVES EXPEDITION

Full appreciation of the Amon Carter Museum's watercolors requires an understanding not only of Möllhausen's artistic and philosophic background but also of the Ives expedition itself. Like Möllhausen's biography, the story of the expedition has been recounted numerous times—usually in sources not easily accessible to the American public—but never in the context of his watercolors. The chief works in English all were written before the rediscovery of the artist's original watercolors. These works and David Miller's recent translation of Möllhausen's expedition diary provide a context for the Colorado River watercolors and related prints.[77]

Möllhausen seems to have pursued his artistic training and scholarly activities after the Whipple expedition with the hope of further travels. Humboldt supported this ambition in a letter dated March 24, 1857, to the outgoing American Secretary of War, Jefferson Davis. Recommending Möllhausen for work as a draftsman or collector of natural history specimens with another U.S. government-sponsored survey, Humboldt wrote:

> Mr. Möllhausen, as an inmate of my house for several months has enlisted my warmest interest in his behalf by his varied knowledge and amiable character. His artistic talents have been singularly developed and perfected here by his intimate relations with one of our first landscape painters. He has read before the Geographical Society in Berlin several memoirs illustrated by his own drawings which have attracted great notice among my friends.[78]

After recounting how the King of Prussia had "manifested a personal interest in Mr. Möllhausen," Humboldt added that "Mr. Möllhausen dreams of nothing but of the happiness to be attached once more to an American expedition."

Through an amazing bit of serendipity, Humboldt's letter received an unexpectedly prompt and favorable reply from Joseph Christmas Ives (1828-1868), a lieutenant in the U.S. Army's Corps of Topographical Engineers (fig. 44). Ives, who had studied at Yale College and graduated from the U.S. Military Academy at West Point, had met Möllhausen while serving as Whipple's second-in-command during the railroad survey expedition. After this expedition, Ives had married Cora Semmes, daughter of the U.S. naval

In 1541 a Spanish expedition under Hernando de Alarcón, attempting to link up with Coronado, discovered the river's mouth and sailed several miles up the Colorado before turning back. Between 1768 and 1776 the remarkable Spanish explorer Father Francisco Garcés explored large segments of the Colorado River, including the canyonlands and plateau country. In 1780 the Spanish briefly established a settlement at the junction of the Gila and Colorado Rivers, but the Yuma Indians soon destroyed it.[82] By the 1820s and 1830s Mexican traders and American mountain men were also familiar with sections of the river.

The United States' war with Mexico and the subsequent acquisition of vast territories surrounding the lower Colorado involved the U.S. government in the river's history. The river was a major barrier for overland travellers to the California gold fields in 1849. To protect them, the army erected Camp Independence, an outpost near the junction of the Gila and Colorado Rivers, in 1850 and formally established Fort Yuma in the same vicinity in March 1851.[83] Supplying the fort became a major problem, so the government sent out several expeditions to explore better routes, including one by water via the Gulf of California and the river itself. In 1851, Captain Lorenzo Sitgreaves led his army topographical expedition west from the Pueblo of Zuñi in the New Mexico Territory to the Colorado River near present-day Bullhead City, Arizona. Contending with low supplies and hostile Indians, he and Richard Kern mapped the left bank of the Colorado River from what is now Davis Dam, on the Arizona-Nevada border, south past present-day Yuma.[84]

During the 1850s, steamboats regularly plied the lower Colorado as far as Fort Yuma, but until 1857 white men still knew little about the river above the fort, despite the maps produced by Sitgreaves' and Whipple's expeditions. The Colorado River's magnificent canyonlands were virtually *terra incognita* for non-Indians, and only a few fragmentary reports provided direct knowledge of a great canyon. The first was from a Spanish expedition commanded by one of Coronado's lieutenants, García López de Cárdenas, who spent three days in 1540 trying unsuccessfully to reach the river from the plateau above.[85] In 1776 Father Garcés also skirted the Grand Canyon's southern rim on a journey from the lower river to the Hopi pueblos. An American, James Ohio Pattie, reported visiting the Grand Canyon in 1826, but his descriptions, published in 1831, were decidedly unscientific.[86]

commander and future Confederate blockade runner Raphael Semmes and niece of the new Secretary of War John B. Floyd.[79] No doubt counting on Möllhausen's past experience, training, and preparation, Ives asked Möllhausen to accompany him as "artist and collector in natural history" on an expedition to survey and determine the navigability of the Colorado River.[80] During the winter months, when the river was at its lowest stage, Ives and his men would ascend the Colorado in a steamboat to determine the head of navigation, and along the way map the river and conduct a hydrographic survey. After reaching the head of navigation, they were to explore by land and continue to map the "vicinity of the river as high up as may be practicable."[81]

Both Ives and Möllhausen had already explored parts of the Colorado River as members of the Whipple expedition, which had reached the river from the east by following Bill Williams' Fork, then had traveled north along the riverbank for several miles before rafting across near the present town of Needles, California, on their way to San Diego, California. Other Europeans and European-Americans had also traversed the lower Colorado River.

An enterprising American civilian and steamboat pioneer named George Alonzo Johnson (1824-1903) contributed to the increased American interest in the Colorado River in the 1850s. Johnson introduced the first shallow-draft steamboat on the lower Colorado in 1852, and for several years his boats had a virtual monopoly on government supply contracts for Fort Yuma.[87] Around 1856 he wrote then-Secretary of War Davis to request funding for river exploration above Fort Yuma. Johnson expected to play a major part in such exploration, but the War Department eventually asked Congress to fund a survey by the army's Corps of Topographical Engineers. Johnson would later become one of the project's sharpest critics, charging the Secretary of War with nepotism for selecting a nephew as its leader.

Ives, however, was a logical choice who proved himself by his performance. With a budget of $40,000 (later raised to $42,000) for the entire expedition, Ives was instructed to, among other things, hire civilian assistants, purchase supplies, and rent or purchase a steamboat. He asked Johnson to quote a rental figure, but Johnson wanted $3,500 per month to use an older steamboat, the *General Jesup*, and $4,500 per month for the newer *Colorado*. For only $8,000—roughly two months' rental—Ives could have a steamboat custom-built, ship it west in parts, and reassemble it at the mouth of the Colorado River. The Philadelphia shipbuilding firm of Reany, Neafie & Company completed a shallow-draft boat with an iron hull and a forty-five-foot-long keel, and by the end of August 1857, the *U.S.S. Explorer* had passed its trials on the Delaware River and was ready to disassemble and ship to the West Coast.[88]

Ives' expedition gained importance with the outbreak of the so-called Mormon War of 1857-58. On May 26, 1857, President Buchanan attempted to replace the Mormon governor of the Utah Territory, Brigham Young, whom the American press had branded a rebel against standard morality and accused of treasonous activities against the government. When Buchanan ordered 2,500 federal troops to march to Utah and enforce his decision, Young called for war. He threatened to burn or destroy everything in the path of the advancing federal troops to prevent supplies from reaching them, and the Mormons would ultimately implement these guerrilla tactics with considerable success.[89] Faced with the difficulties of getting men and supplies to this remote territory, the army saw the Colorado River as a possible alternate supply route, if

steamboats could go far enough upriver to land near the Old Spanish Trail into Utah. Moreover, an additional U.S. Army presence on the Colorado River would help cut off any Mormon exodus to Sonora.[90]

Möllhausen evidently had little or no knowledge of the steamboat, the unfolding Mormon situation, or of Johnson when he accepted Ives' invitation in June 1857.[91] The Whipple expedition had made him aware of many potential dangers, however, and he took the position in full knowledge that he would have to leave his wife and child for an extended period. Surviving documents do not define Möllhausen's duties exactly, but as the expedition's civilian artist, he was to record as precisely as possible the landforms, vegetation, animals, indigenous peoples, and settlements they encountered. His new position was as "artist and collector in natural history" rather than "topographer and draughtsman" (as he had been with Whipple). The expedition's topographer, another civilian, was to concentrate on mapmaking but would also contribute pictorial material to the final report, while Möllhausen was to collect zoological specimens and assist the expedition's geologist by sketching formations.

Before he left Prussia, Möllhausen wrote Smithsonian Director Spencer Baird about his arrangements for collecting natural history specimens, noting also that he had been preparing himself "in the *ateliers* of our best artists and conservators . . . in the hope to gain some credit by my work."[92] He visited libraries and collections in Prussia and later in Washington, D.C., to study illustrated publications on America, including the recent United States and Mexican Boundary Survey reports of Major William H. Emory and the report of a private railroad survey expedition led by A. B. Gray.[93] In August, just before leaving for America, Möllhausen had another audience with King Friedrich Wilhelm, who on Humboldt's recommendation rewarded Möllhausen's earlier presentations with fifty gold pieces and "the Order of the Red Eagle, fourth class." Humboldt advised Möllhausen to speak of these awards "with no one except your wife, in-laws, and family," because "envy flourishes in Berlin among artists and employees of the royal court"; instead, the public could learn about them after Möllhausen was away in the United States.[94]

He left for the United States on August 12, 1857, met with Ives after he arrived in New York, then proceeded to Washington, D.C. The number of influential people he visited there attests to the perceived importance of the expedition. In addition to old friends and acquaintances such as Professor Spencer Baird of the Smithsonian and Baron Leo von Gerolt, the Prussian ambassador, Möllhausen had an interview with Secretary of War Floyd, and, through the invitation of the prominent American banker, philanthropist, and art collector William Wilson Corcoran (1798-1888), paid a visit "in the country" to President Buchanan. The German noted that both the President and Secretary of War "encouraged him about the impending expedition."[95] As with any government-sponsored operation, the political intrigue must have been considerable, and not everyone in Washington or in Prussia approved of Möllhausen's influence or his art. Evidently the news of Friedrich Wilhelm's generosity to Möllhausen prompted some people to urge that the American government withdraw its support. Möllhausen showed Secretary Floyd samples of his latest artwork to prove that he was capable of better work than "the horrible prints" that had appeared in the Whipple report, and to Spencer Baird he expressed the hope that the pictures would "put every thing in order, and make ashamed the German friends which took a pleasure in slandering me or in trying to put down my good will."[96]

On September 21, 1857, Möllhausen departed from New York for the West Coast by way of Panama. His travelling companions on this long voyage were Dr. John Strong Newberry (1822-1892) and Baron Friedrich W. von Egloffstein (1824-1885). Newberry (fig. 45), a professor of geology at Columbian (now George Washington) University and a Smithsonian member who had already served on a railroad survey expedition in 1855,[97] was to serve as physician and geologist to the expedition and to "take charge of the natural history department," with Möllhausen to assist him. Egloffstein (fig. 46), the expedition's topographer, was a Bavarian nobleman with a penchant for adventure, who had served previously with exploring expeditions under John C. Frémont and Lieutenant E. G. Beckwith in 1853-54.[98] A very capable and experienced topographer and mapmaker, Egloffstein was to make topographical sketches for Lieutenant Ives—a task that would overlap Möllhausen's own work to some extent.

On October 22, 1857, Möllhausen, Newberry, and Egloffstein arrived in San Francisco, where they were greeted by Ives and other expedition members who had preceded them there on a different ship. Ives divided the expedition into three groups. Newberry's group was to go by coastal steamer to San Diego, then overland to

Fig. 45
John Strong Newberry, photograph courtesy of the Smithsonian Institution

Fig. 46
Baron F. W. von Egloffstein, 1865, photograph courtesy of the Kansas State Historical Society

Fort Yuma. The group that included Möllhausen and Egloffstein was to travel by coastal steamer to San Pedro, side-trek overland via Los Angeles and Mission San Fernando to Fort Tejon to procure mules for the entire expedition, then return to Los Angeles and journey overland to Fort Yuma. Meanwhile, Ives' contingent (including A. J. Carroll, a Philadelphia engineer and the master mechanic for the *Explorer*) would sail around the Baja peninsula with the disassembled steamboat on board the schooner *Monterey*; at the mouth of the Colorado, they would put the *Explorer* back together and steam upriver to rendezvous with the other two groups at Fort Yuma.

Möllhausen had begun keeping a voluminous diary when he left for the United States, and along with Ives' official *Report* and his own sketches and watercolors, it provides much of the documentation of the Ives expedition. Eventually published in Germany as *Reisen in die Felsengebirge Nordamerikas . . .* (1861), it covers not only the upriver exploration and the overland journey into the Grand Canyon but also Möllhausen's activities until his return to Europe in August 1858. Like his earlier diary of the Whipple expedition, *Reisen* vividly portrays the landscape, flora, fauna, people, customs, events, and incidents in the American Southwest of the 1850s.

Moreover, with more experiences behind him, Möllhausen's diary exhibited greater accuracy, discernment, and insight in discussing these subjects. While his style of prose may be verbose and overly romantic, it is seldom dull or boring, even though he carefully recorded both important events and all kinds of associated minutiae. Often his descriptions are so vivid that his German audience must have felt as if they were on the expedition themselves. While Ives' *Report upon the Colorado River of the West . . .* (Washington, 1861) provides the official account of the expedition (including Dr. John Strong Newberry's geological report), Möllhausen's *Reisen* provides much more information about daily life on the expedition. Moreover, as a foreigner writing for a foreign audience, Möllhausen obviously had a different perspective from Ives on many issues, and at times he wrote critically of his commander, the U.S. government, and Americans in general, being reasonably confident that his words would probably never reach the attention of most of his American associates.

The diary offers many interesting details about his journey across the Isthmus of Panama and his stay in California, where he saw Grizzly Adams' tamed bears on exhibit in San Francisco, visited with General Andrés Pico at Mission San Fernando, and reported on other California pioneers. He also began making a sketchbook of small graphite drawings, now with his descendants in Germany.[99] His field studies of landscapes, people, plants, animals, and camp scenes—with a geographical range from California, along the Colorado River to the Colorado Plateau, through the present-day states of Arizona, New Mexico, and Kansas—sometimes have little to do with the official business of the expedition. A number are genre scenes, likely intended for the artist's private purposes, and many of the sketches pertain to geographical areas that were not officially within the expedition's scope (particularly the New Mexico and Kansas views). Significantly, there are no color notes on the surviving field sketches, and only one has watercolor as well as graphite.

Although Möllhausen began to sketch before he arrived at Fort Yuma, it is not clear which of the surviving images he produced first.[100] One of his earliest, apparently, depicts one man on a mule, frightened by a camel (see cat. 26b). If, as his realistic portrayal of the camel suggests, he relied on direct observation from nature, he must have made the sketch in California, where Möllhausen saw camels in Los Angeles on November 9, 1857, and shortly afterwards

at Bishop's Ranch near Fort Tejon.[101] The exotic dromedaries, part of an experiment in desert transportation conducted by the War Department, would later figure prominently in one of Möllhausen's watercolors (see cat. no. 26).

Perhaps most interesting from an artistic standpoint is the continuing improvement of Möllhausen's drawing skills in these sketches. The proportions of the animals, people, and vegetation show a direct and faithful observation of nature. The artist applied various amounts of pressure with his pencil strokes—darker in the foreground and fading back as objects recede in the distance—and for the first time, he used regularized shading and cross-hatching with parallel diagonal lines, employing swift strokes that attest to a practiced hand.

Unlike the camel sketch, most of the drawings in Möllhausen's sketchbook do not relate to his later watercolors. He titled and dated many of his sketches, but a few are more problematic. A landscape titled *Colorado Ufer* ("Bank of the Colorado") does not offer enough clues to pinpoint where it was sketched, and a few other landscapes, some studies of trees, and a sketch of a hound and a rabbit likewise cannot be identified exactly even after a search of the artist's diary.

Presumably a number of Möllhausen's original field sketches have not survived. In *Reisen*, for example, he wrote of having to make "numerous and large drawings . . . on the Colorado"—clearly not a reference to the small sketchbook. He also mentioned having watercolors with him and specifically stated that he both *painted* and sketched while on the expedition.[102] However, the forty-six watercolors in the Amon Carter Museum and those few surviving watercolors relating to the Ives expedition in the Berlin Museum of Ethnology are clearly not field sketches (the Berlin Museum's other Möllhausen watercolors were destroyed during World War II and now are known only through photographs).

The extant watercolors' fine condition and finished quality of composition and rendering suggests that Möllhausen did not draw them in the wilderness and then carry them with him on the *Explorer* or transport them hundreds of miles on muleback through rivers and gulleys and over deserts, mountains, and plains. Nor does his journal mention that he worked up any of his Ives expedition field sketches into finished watercolors during his return journey to Washington, D.C. The subjects of the surviving watercolors can be identified, dated, and clarified through his detailed descriptions in his personal journal, but the dates apply only to the original field sketches from which he must have prepared the finished watercolors.

The rediscovery of Möllhausen's watercolor originals aids in distinguishing his contributions to Ives' *Report* from the work of others, although it does not solve the problem entirely. Ambiguous wording in the report's text generally attributes the images to either Möllhausen or Egloffstein, and while all of the lithographs credit the artist, the forty-one woodcuts in Ives' account are simply "Drawn by Mr. J. J. Young from sketches by Messrs. Mollhausen and Egloffstein," and the twenty-seven woodcuts in Newberry's geological report give no credits at all. None of Egloffstein's original sketches for the Ives expedition are known to exist today, but an original Egloffstein sketch from his earlier travels with Beckwith reveals him to have been a very careful and precise draftsman in his delineation of the landscape. Dated and captioned "April 5th at 2 P.M. from an Island in Weber River, Valley of Great Salt Lake," the sketch (fig. 47) gives an indication of Egloffstein's meticulous working methods and served as the model for an engraving titled *Weber Lower Cañon* in volume eleven of the U.S. Pacific railroad survey reports. Although none of the sources mention Newberry making sketches while with the Ives expedition, the geologist may also have provided some, for he was a capable artist who later furnished sketches for Captain John N. Macomb's expedition report in 1859.[103]

Before the rendezvous at Fort Yuma, Lieutenant Ives himself began producing visual images. At the mouth of the Colorado River in early December 1857, and also later, he made small topographical views in his official fieldbooks of the expedition, as well as topographical notations and well-drawn map plats. Ives had undoubtedly studied topographical drawing at West Point, but his views are generally very crude and do not show any exceptional artistic skill. One of their most interesting features is the holes burned in the paper of some by falling cinders from the *Explorer*'s smokestack.[104]

Nonetheless, one of the first lithographs in Ives' *Report* credits "J. J. Young from a photograph by Lieut. Ives" (fig. 48). The print shows the schooner *Monterey*, which Ives used to transport the disassembled steamboat *Explorer*, docked at Robinson's Landing, a few miles up from the Colorado River's mouth. Just to the right of the schooner stands a small house on stilts belonging to David C.

FIG. 47
F. W. von Egloffstein, *Lower Weber Canon, North of Great Salt Lake City*, ink wash on paper, 1856, Amon Carter Museum

FIG. 48
John James Young after a photograph by Joseph C. Ives, *Robinson's Landing, Mouth of Colorado River*, toned lithograph (black with ruled gray), from Ives, *Report* (1861), part 1, opp. p. 27, Amon Carter Museum Library

Robinson, an employee of George A. Johnson and himself an experienced riverboat captain, whom Ives hired to pilot the *Explorer*. The little steamboat, although scarcely visible in the print, is still being assembled and rests in a deep pit to the left of the schooner. In his report, Ives explained the origin of the image:

> There being a little photographic apparatus along, I have taken advantage of the mild and quiet interval to experiment, and having constructed out of an india-rubber tarpaulin a tent that entirely excluded the light, have made repeated efforts to obtain a view of camp and the river. The attempt has not met with distinguished success. The chemicals seem to have deteriorated, and apart from this the light is so glaring, and the agitation of the atmosphere near the surface of the ground so great, that it is doubtful whether, under any circumstances, a clear and perfect picture could be secured.[105]

Ives found the difficulties of early photography to be insurmountable, and a few days later a gust of wind blew away the photography

FIG. 49
After Balduin Möllhausen, *Cocopas*, chromolithograph
(black with red, gold, and blue), from Ives, *Report* (1861), part 1,
opp. p. 31, Amon Carter Museum Library

tent, "apparatus and all."[106] Unfortunately, none of his original photographs from the expedition, including the one from which the lithograph was derived, is known to survive. Significantly, though, he had anticipated a major shift: within a few years photography would supplant sketches and paintings for scientific illustration.

The reassembly of the *Explorer* took three weeks—much longer than anticipated. Ives and his men had to contend with the rising and lowering of the tide or bore of the river, which submerged nearly all of the land around Robinson's Landing at high tide and flood season, and with the shifting sands of the river's delta. Another problem was the steamboat itself. It had been designed with an oversize boiler, but the hull had not been strengthened enough to support the extra power; thus the vessel would vibrate and shimmy when in operation. To rectify the problem, Carroll, the engineer, bolted several planks to the boat's underside, but this solution would severely limit the expedition: since the planks created more drag than a smooth hull, the steamboat frequently grounded in the shoals and shallows of the river.[107]

Meanwhile, Möllhausen had arrived at Fort Yuma with his overland party on December 20, and during their three-week wait for Ives, he sketched the scenery and the local Yuma (Quechan) and Cocopa Indians who frequented the fort.[108] One of his sketches may have served as the model for the lithograph titled *Cocopas* (fig. 49), showing the Native Americans who inhabited the Colorado River delta below Yuma.[109] Möllhausen's print depicts two Cocopas wearing what appear to be cotton trade clothes. The man at left has Native American features, and both wear their hair at a typical length with slender headbands around their foreheads. One has on sandals, and the other seems to wear a bead necklace of some sort. A bow and arrows with quiver, which the Cocopa used for hunting or war, hangs from a branch of a leafless tree. Their appearance may be atypical for that time, however. Lieutenant Ives, who had several opportunities to observe the Cocopa, described the men's clothing as "a narrow strip of cotton, tied about the loins, an unseasonable dress for this time of year," and considered the women's apparel "about as deficient . . . a very short petticoat, their only garment, taking the place of the strip of cotton."[110] After another visit from the Cocopas, the lieutenant noted:

Fɪɢ. 50
John James Young after Balduin Möllhausen, *Yumas*, chromolithograph
(black with blue, gold, and red), from Ives, *Report* (1861),
part 1, opp. p. 44, Amon Carter Museum Library

A few are provided with blankets, but nearly all, males and females, are on a scanty allowance of clothing. The women generally have modest manners, and many are good looking. They have a custom of plastering their hair and scalps with the soft blue clay from the river bank, the effect of which is not at all pretty, but the clay is said to be a thorough exterminator of vermin, and as such must give them a great deal of comfort.[111]

Yumas (fig. 50), another lithograph in Ives' *Report*, also seems to have been made from Möllhausen's sketches. By the time of Ives' expedition, the Yuma—whose lands lay at one of the most strategic crossing points on the Colorado River—had had considerable experience with Europeans, from Spanish explorers and missionaries in the late eighteenth century to hordes of American forty-niners and settlers in the mid-nineteenth.[112] Möllhausen had first met some Yumas, or "Cutchanas" as he called them then, in late February 1854 with the Whipple expedition. At the time he had noted that they were a "vigorous, powerful race, amongst

whom a man of less than six feet high appeared to be quite a rarity," although the women tended to be "short" and "thickset." Both sexes, he reported, "wore the hair cut short over the eyebrows, but the women never have their hair twisted into tails." He also observed that the men carried a short war club or mallet as well as bows and arrows, and thus were called "Club Indians," or "Galloteros."[113]

Although nobody recorded the identities of the three Yuma men in the lithograph after Möllhausen's sketch, the one in the middle, wearing a red shirt, may have been Pasqual II (d. 1887); this figure answers descriptions of Pasqual and bears a likeness to a photograph made of the chief in the late 1880s.[114] A well-known Yuma *capitan* or *kwoxot* who was noted for his friendliness to Americans, Pasqual visited Möllhausen's camp the very day the explorers reached the Colorado, to present letters of recommendation from various officers at Fort Yuma. Another man in the print may be Maruatscha, an adventurous Yuma *capitan* who served the Ives expedition as an interpreter and travelled with them on their steamboat until February 20.[115] The third could have been a "gigantic Yuma" who hitched a short ride with them on January 18.[116] In the lithograph all three figures wear various clothes of European, American, or Mexican manufacture, although an 1878 source noted that Pasqual only wore pantaloons when he expected to interview or be interviewed by whites, and not always then.[117] Möllhausen may not have seen the three men at once but could have sketched them individually; until January 21, when the expedition passed upriver out of their territory, the artist had several opportunities to observe and sketch members of the Yuma tribe.

While waiting for Ives at Fort Yuma, Möllhausen eagerly listened to any news from the outside world, noting that rumors were circulating there about impending hostilities with the Mormons. Of special concern were reports about Mormon agents actively recruiting among the Mohaves farther upriver. At the same time, Ives' rival, the steamboat operator Johnson—who had been observing the expedition's progress and had been planning one of his own—saw the rumor of Mormon agents as an opportunity to upstage Ives. Johnson had conversed with Ives on December 17, while taking on cargo from the *Monterey* for transport back to Fort Yuma, and had told the lieutenant that in three days he was going upriver with the steamer *General Jesup*. Johnson had added: "My vessel is large enough to take you and your steamer on board, and not notice it. If I find the river navigable, I will have it published to the world before you can launch your boat and leave tide-water."[118] Rumors of Mormon activity convinced Fort Yuma's acting commander to send an escort of troops, a howitzer, and military supplies upriver on the *General Jesup*, giving the expedition semi-official status as Johnson tried to accomplish his personal goal of navigating the Colorado River.

Although Ives' official *Report* omitted nearly all reference to Johnson's expedition,[119] Möllhausen wrote candidly that he and his group watched with dampened morale as Johnson, accompanied by a fourteen-man military escort under Lieutenant James L. White, set off in the *Jesup* on December 31. "Naturally such a step, undertaken in greatest haste just prior to the arrival of Lieutenant Ives and our own steamboat aroused indignation and mistrust," and the *Jesup* "did not exactly take our warmest blessings with her."[120] Möllhausen speculated that such behavior might have arisen from the "jealousy between the officers of the line and those of the Corps of Topographical Engineers," or perhaps the "speculative fever of certain persons" who hoped to find gold or silver deposits along the Colorado River.

In actuality, Johnson's hasty trip up the Colorado may have benefitted Ives. On December 22, while at Robinson's Landing, Ives received two letters from his superior in Washington, apprising him of the Mormon situation and directing him to make a preliminary reconnaissance of the river before beginning his regular survey, to determine "the practicability of and facilities for sending large bodies of troops from Fort Yuma to Great Salt Lake along the Colorado and Virgin Rivers."[121] For the immediate purposes of the War Department, the Johnson-White expedition could serve as the preliminary survey, and Ives could continue the more thorough regular survey at a pace better suited to scientific accuracy.

Ives and the *Explorer* finally arrived at Fort Yuma on January 9, 1858, joining the two overland groups already there. He busily issued his final instructions and oversaw the loading of the expedition's supplies and equipment on board the *Explorer*. As space was limited on the small vessel, he ordered the two topographical and meteorological assistants, Paul H. Taylor and Charles K. Booker, to remain behind at Fort Yuma and join the expedition later. He also hired an experienced California herder, G. H. Peacock, to assemble a mule train that would resupply the expe-

dition with fresh provisions from Fort Yuma after the steamboat had reached the head of navigation.

The expedition members and the garrison at Fort Yuma celebrated the night before the *Explorer* departed. Möllhausen recorded that they drank liquor and Rhine wine, sang, and danced; he and Ives accompanied on the guitar, with Egloffstein and Carroll on flutes and Newberry on the violin. Dancing followed the singing as bearded men embraced one another and stamped "with heavy heels to our crude music of the eternal Yankee Doodle and the Scottish Horn Pipe." One of the garrison members toasted the explorers: "Think of us when the Mohave Indians lift your scalps!" The explorers responded: "Remember us when the Mormons burn you and your fort to the ground."[122]

They left early the next day, January 11, 1858, with their two Indian interpreters, Maruatscha and Mariano, aboard the *Explorer* and a group of Yuma men, women, and children gathered to see them off:

> The men grinned, and the women and children shouted with laughter, which was responded to by a scream from the Explorer's whistle; and in the midst of the uproar the line was cast off, the engine put in motion, and, gliding away from the wharf, we soon passed through the gorge abreast of the fort and emerged into the open valley above.[123]

Foreshadowing the troubles that lay ahead, the little steamboat almost immediately grounded on a sandbar and remained there most of the day before her crew could get her off again. Ives added:

> The delay would have been less annoying if it had occurred a little higher up. We were in plain sight of the fort, and knew that this sudden check to our progress was affording an evening of great entertainment to those in and out of the garrison.

Innumerable times during the course of the trip, they would encounter such delays, which often involved unloading passengers and supplies. However, these episodes gave Möllhausen more time to sketch, write, and collect natural history specimens. He and Newberry often conducted scientific excursions of their own away from the rest of the group—a practice that was not without danger in a land of hostile Mormons, Indians, and wild, somtimes poisonous animals. Likewise during these delays, Baron von Egloffstein would often wander off into the surrounding hills, sometimes by himself, in order to make a topographical sketch from an elevated position. This predilection for rock climbing may have been natural for the baron, whose family's lands were located in an area of Bavaria known as "the Franconian Switzerland" for its picturesque hills.[124]

The incongruous aspects of the expedition invited humor for both its members and the native peoples who observed them. Möllhausen compared the *Explorer*'s ungainly appearance to a wheelbarrow, and the Cocopas called it the "chiquito steamboat" because it was so much smaller than Johnson's vessels.[125] Captain Robinson learned to anticipate shallows by watching Yumas collect on the banks at points where they expected the steamboat to run aground.[126] Noting how the peaceful Chemehuevi Indians always seemed to be laughing, Ives learned that they thought the white men were incredibly stupid and that the whole idea of a steamboat was ludicrous to them—especially considering that in much less time they could walk, swim, or raft the same distance covered by the iron vessel. "The gleeful consciousness of superiority at all events keeps them in an excellent humor," the lieutenant wrote.[127] Möllhausen was also a source of amusement to the Chemehuevi. He gave the children strings of beads to bring him specimens of rats, mice, lizards and snakes, and Ives noted: "They think he eats them and are delighted that his eccentric appetite can be gratified with so much ease and profit to themselves."[128]

The steamer's smokestack spewed burning embers, which landed on the after-deck and the platform over the wheelhouse, where Möllhausen, Egloffstein, Newberry, Ives, and Casimir Bielawski, the hydrographer, often sat. The cinders burned through their hats, coats, and even their notebooks, and this frequently led to humorous scenes, such as the one described by Möllhausen:

"Someone's on fire!" yells one of the men, whose nose first detects the aroma of singed wool. Everyone jumps to his feet and examines himself from head to toe and on all sides. The bad odor becomes pungent as examining eyes look sharper in the folds of the clothing. Suddenly someone yells, with a burst of laughter, to the very person who first sounded the fire alarm. "Your own head's on fire!" His startled hand grabs for

Fig. 51
John James Young after Balduin Möllhausen,
Spire Range, engraving, from Ives, *Report* (1861), part 3, p. 23,
Amon Carter Museum Library

the only headgear that he has, a hat which in the full sense of the word has become topless.[129]

Proceeding upriver at an incredibly sluggish pace, the *Explorer* passed through Purple Hill Pass, Canebrake Canyon, a cluster of graceful spires near Chimney Peak (fig. 51), the Red Rock Gates of the Chocolate Mountains, and the Great Colorado Valley (see cat. nos. 3-11 and map). It continued by degrees past the Half-way (Big Maria) Mountains, the Riverside Mountains, and through Monument Canyon (cat. nos. 13-17). At every turn the scenery became more colorful and rugged, commanding the attention of Möllhausen and Professor Newberry, the geologist, who studied it as Möllhausen sketched exposed layers of rock strata and interesting formations where the river cut through ranges of mountains (fig. 52).[130]

Travelling up the river, the expedition encountered Mohaves, with their distinctive tattoos and grass-skirted women (see cat. no. 12), and Chemehuevis; both sometimes followed them along the

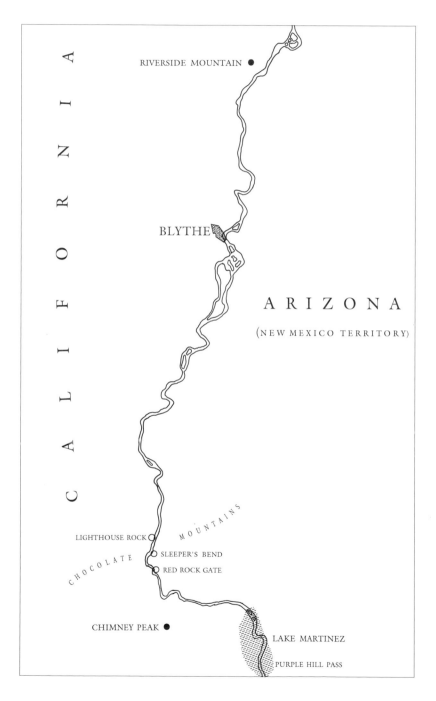

FIG. 52
After Balduin Möllhausen, *Remains of Grand Mesa in
Chemehuevis Valley*, engraving, from Ives, *Report*, part 1, p. 62,
Amon Carter Museum Library

riverbank and visited the explorers' camps to trade and barter.
Möllhausen had been one of the earliest, if not the first European
to sketch the Chemehuevi when the Whipple expedition encountered them in 1854 (fig. 53).[131] When he met them again with the
Ives expedition, he noted that they had "more handsome physiognomies among them than in any other tribe, and I caught sight of
individual profiles which did not exhibit a trace of the Indian type,
but rivaled the regularity of genuine Roman profiles." However,
he noted, "in strength and stature they are significantly inferior
to the Mohaves."[132] Similarly, Lieutenant Ives reported that the
Chemehuevis were

> altogether different in appearance and character from the
> other Colorado Indians. They have small figures, and some
> of them delicate, nicely-cut features, with little of the Indian
> physiognomy. Unlike their neighbors—who, though warlike,

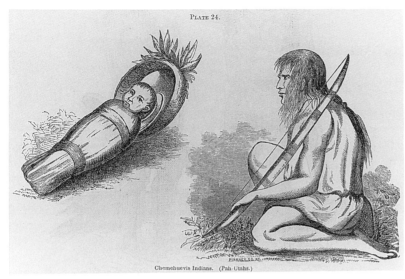

PLATE 24.

Chemehuevis Indians. (Pah-Utahs.)

FIG. 53
After Balduin Möllhausen, *Chemehuevis Indians. (Pah-Utahs.)*,
engraving, from *Reports of Explorations and Surveys* (Washington, 1856),
vol. 3, part 3, p. 32, Amon Carter Museum Library

arc domestic, and seldom leave their own valleys—the Chemehueris [*sic*] are a wandering race, and travel great distances on hunting and predatory excursions. They wear sandals and hunting shirts of buckskin, and carry tastefully-made quivers of the same material.[133]

However, in his diary entry on January 24, Möllhausen observed:

The Chemehuevis who visited us today seemed to live in want. Families were reduced to a level of poverty, and the women and children were unclean. They even lacked articles of Indian clothing. . . . Only the chief of this little band was decked out in bizarre finery, and it appeared to me as if he had used the entire wealth of his few subjects to secure a stately appearance for himself. He wore a red woolen shirt, and over this a narrow blue wool shawl. A wide flame-colored strip of cloth encircled his head and a thick plume of owl feathers was fixed in his hair. He had hung a veritable ballast of white

porcelain beads around his neck. These emphasized the strident colors of his fantastic attire. The old wrinkled warrior did not look too bad. He was well aware of this, although his comrades did not even glance at him, much less speak to him. Since our people did not accord him the deference and admiration which he probably expected, he appeared to be very offended.[134]

Möllhausen apparently sketched this personage, carrying a standard bow; he is the central figure of a chromolithograph in Ives' *Report* (fig. 54). The warrior at right, wearing a headband and holding a smaller, double-curved mountain sheep horn bow, has a trade knife secured to his waist and a leather quiver strapped around his shoulder, while the woman at left wears a traditional bark skirt and cape and holds a child in a cradle.[135]

In exchange for such items as beads and blankets, Indian runners kept the expedition in touch with Fort Yuma.[136] Similar products served as pay for the expedition's interpreters, a Spanish-speaking Diegeño named Mariano and the young Quechan *capitan*

FIG. 54
John James Young after Balduin Möllhausen, *Chemehuevis*,
chromolithograph (black with blue, red, and gold), from Ives, *Report*
(1861), part 1, opp. p. 54, Amon Carter Museum Library

named Maruatscha.[137] Communication in the far west could be "a complicated process," as Ives admitted understatedly. At times the lieutenant had to deliver a message to Bielawski, the Polish-American hydrographer from San Francisco, who put it "into indifferent Spanish" for the Diegeño interpreter, Mariano, whose knowledge of that language was "slight."[138] Mariano would then communicate with Maruatscha, the Quechan. They in turn must have felt themselves lucky if they had to communicate with the Mohaves, who could at least understand some of the Quechan language because it was similar to their own; the Chemehuevi, on the other hand, were linguistically more related to the Paiute.

While Ives struggled to communicate with the various Native American groups and the *Explorer* worked its way upriver, Captain Johnson and Lieutenant White, on board the *General Jesup*, had already steamed to a point near the modern border of Arizona and Nevada (today inundated by the waters of Lake Mohave), where they had been deterred by rapids. Lacking a cable to winch the *Jesup* over them, Johnson instead erected a rock cairn at the site and called it the head of navigation. (Ives would completely ignore this in his *Report*, but Möllhausen documented Johnson's "monument" in one of his watercolors [cat. no. 28]). On January 23, on their way back downstream near present-day Riviera, Arizona, Johnson and White chanced upon Lieutenant Edward Fitzgerald Beale's survey expedition. Beale and his men—accompanied by camels, mules, and wagons—had come from Los Angeles and Fort Tejon, surveying for a wagon road to connect the Colorado River with Fort Defiance on the Arizona-New Mexico border. Johnson and his steamboat ferried Beale's party across the river before the two expeditions parted company, and this coincidental encounter produced a big impression upon everyone present, including White, Johnson, and the Mohaves (see cat. no. 26).

Continuing downriver, the *Jesup* met the rival *Explorer* steaming upriver just below Monument Canyon on January 30. After members of the two expeditions conversed with one another and exchanged information, Ives dismissed the commander of his escort, Lieutenant Tipton, who was to accompany the *Jesup* back to Fort Yuma. With this action, Ives made Tipton available to accompany the mule train of supplies that was to meet the *Explorer* after it had gone upriver as far as it could. Unfortunately, the *Jesup* struck a rock and sank thirty miles from Fort Yuma. While no one on board was seriously injured, Johnson, White, and their men had to walk the rest of the distance back to the fort. Given the rivalry between the two expeditions, Ives and his men must have heard this news with some satisfaction when they learned of it almost a month later. White would eventually file an official report of his expedition, but it was only ten pages long, with one map—and thus of little lasting importance when compared with the elaborate reports, sketches, and maps of Ives and his better-qualified team of experts.

When the expedition reached the mouth of Bill Williams River on February 1, Möllhausen and Ives recognized, with some difficulty, scenery already familiar to them from the Whipple expedition (see cat. nos. 19-20). The *Explorer* slowly progressed through the Chemehuevi Valley and entered quiet and majestic Mohave Canyon, formed where the river cuts through the Mohave Range near its unusual pinnacles, which Whipple had called "the Needles" (see cat. nos. 21-23 and map, p. 43).

On February 10 the steamboat emerged from the other end of the canyon into the Mohave Valley. For the next several days, crowds of Mohave men, women, and children flocked down to the river from their nearby villages to see the expedition and to trade. Two old acquaintances of Ives and Möllhausen, chief Cairook and the warrior Iretéba, appeared and agreed to serve as guides, as they had earlier for Whipple's expedition. Their presence helped to insure that the expedition's relations with the Mohaves remained peaceful, and once the two men boarded the *Explorer*, along with one of Cairook's four wives, Ives' expedition resumed its journey upriver.

The tall Mohave men impressed Möllhausen and Ives just as they had earlier during the Whipple expedition, but the presence now of the beautiful young "Madame Cairook" especially charmed the two white men (see cat. no. 24). Clad in little more than a bark skirt, she watched all the machinery of the steamboat with uninhibited poise and spontaneous enjoyment, and, after a short ride, disembarked to relate her "adventure" to some friends on the shore. Her husband soon followed her, and, in Cairook's absence, Iretéba's friend Navarupe joined the expedition.

Not all of Möllhausen's artistic efforts were committed to paper. The Mohaves, including young girls, were fond of tattoos and body-painting and asked Möllhausen to apply his watercolors to their faces, limbs, and torsos. "It was almost touching when a young mother timidly came up to me and held up her infant for me to paint," wrote Möllhausen.

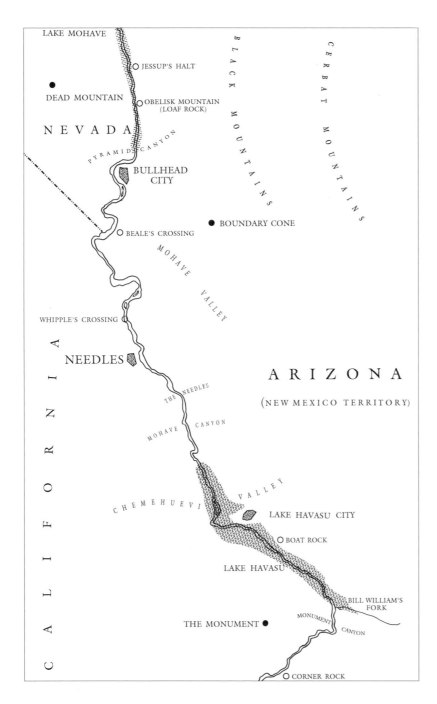

I painted his forehead green, his eyelids yellow, his cheeks blue, his ears and nose red, and his chin violet. As I painted I was pleased by the tenderness with which the mother looked at her child as I transformed it into a little chameleon. When I had finished my work the woman rushed to her companions, where her darling became an object of general amazement.[139]

In his journal Möllhausen recorded generally favorable descriptions of the Mohaves and often attempted to draw a moral lesson about the universality of human nature. Throughout *Reisen*, in comparing so-called primitive peoples with their "civilized" counterparts in Europe and America, he often stressed that all cultures had their advantages and disadvantages, their good and bad sides.

On February 16, the steamboat reached "Beale's Crossing," where Lieutenant Beale's wagons and camels had crossed the river. On board the steamer with Ives and his men at this time was an English-speaking Mohave known as Captain Jack, who had witnessed at least one of Beale's river crossings and who may have helped inspire Möllhausen's subsequent watercolor sketch titled *Beale's Crossing* (cat. no. 26).[140]

The *Explorer* continued slowly on through Pyramid Canyon, above present-day Bullhead City, Nevada, and Davis Dam. With great toil, Captain Robinson and his crew got the steamboat past the rapids where the *Jesup* had turned back, but above them, many more rapids made the journey even more arduous and slow-paced. Möllhausen, Newberry, and Egloffstein took advantage of this to make short excursions into the nearby hills and mountains. By March 4, however, the strains of the trip seemed to show in both Möllhausen's and Ives' written narratives. Both are almost totally silent on their activities of March 4, 1858; although Möllhausen made three field sketches, his journal merely recorded that they "made 2½ more miles with a great deal of difficulty."[141]

He apparently spent a lot of time sketching or engaged in other amusements such as skipping rocks or playing with Grizzly, the dog that had followed them from California. These activities are depicted in *Teasing the Dog* (fig. 55); the man on the right might be the artist or any of a number of expedition members, including Baron Egloffstein, who wore a similar coat and hat in the field sketch called *Four Musicians* (see cat. 8a). The Indian on the left is possibly Iretéba or Navarupe. Both men sit on what appears to be the gunwale of the *Explorer*.[142] In the sketch titled *Amusements on the*

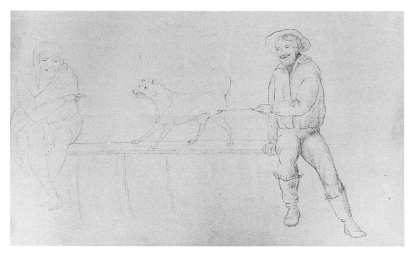

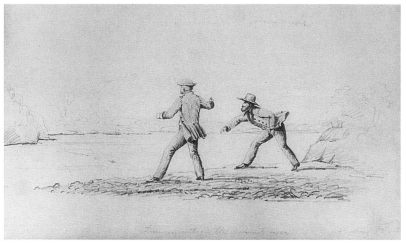

FIG. 55
Balduin Möllhausen, *Teasing the Dog, March 4, 1858*, graphite on paper, from Sketchbook 3, Möllhausen Family, Bleicherode, Germany, photograph courtesy of Dr. Friedrich Schegk, Munich

FIG. 56
Balduin Möllhausen, *Amusements on the Colorado River, March 4, 1858 [Zwei Männer am Flußufer]*, graphite on paper, from Sketchbook 3, Möllhausen Family, Bleicherode, Germany, photograph courtesy of Dr. Friedrich Schegk, Munich

Colorado River (fig. 56), two men skip rocks across the river. The bearded figure in a wide-brimmed hat on the right could be a self-portrait, but the man also resembles the portrait of Dr. Newberry playing a violin in the *Four Musicians* sketch. The man on the left, in a military jacket with insignia patches on the shoulders, is probably Lieutenant Ives.

The relative inactivity of the expedition members during these delays may have resulted in part from their poor physical condition. Lieutenant Tipton's mule train of supplies from Fort Yuma was long overdue, and by February 28 Möllhausen complained that the men had been "reduced to eating beans and maize" as their "only means of sustenance," a situation that was

> especially vexatious because we had no salt. Our people had begun to feel the effects of our poor diet, for not only did we suffer stomach diseases, but scurvy also began to attack the gums of a few and to show up as flecks on the skin. So we placed great value on things we would scarcely have given a thought to under other circumstances.[143]

The sketch titled *Nothing But Beans* (fig. 57) depicts this serious

condition with grim humor, and suggests that by March 4 even their maize was gone. The seated figure may again be a self-portrait, and the double-barreled shotgun may have been the one the artist used for bird hunting.

Ives and his men also began to feel more vulnerable to danger in the vast, lonely wilderness. They had entered a territory frequented by hostile Paiutes and virtually unexplored by white men. With provisions running low and the supply train still a long way downriver, they had to depend upon the Mohaves for much of their food and for any communication with the pack train and Fort Yuma. Ives and his men kept continuous watch for Mormons, who, as the explorers learned from their Mohave friends, had been among their tribe and among the Paiutes, attempting to prejudice them against the Americans. In fact, Mormon missionaries had tried to convert the Mohaves and had baptized several, including Cairook and Iretéba. However, despite (or perhaps because of) these conversions, these two guides demonstrated good intentions in all of their dealings with Ives' expedition. Although placed in an awkward position between the two rival white factions, their main goal was evidently to maintain a peaceful neutrality.

FIG. 57
Balduin Möllhausen, *Nothing But Beans, March 4, 1858,* graphite on paper,
from Sketchbook 3, Möllhausen Family, Bleicherode, Germany,
photograph courtesy of Dr. Friedrich Schcgk, Munich

The *Explorer*'s voyage upriver ended abruptly at the southern entrance to Black Canyon (see cat. no. 33). There, on March 6, the steamboat struck a submerged rock with a violent jolt that threw one man overboard and everyone else forward onto the deck. Fortunately, no one was seriously injured, and the *Explorer,* though damaged, stayed afloat, allowing Möllhausen and the others to save their notes and sketches. Ives decided to halt the expedition there and wait for the mule train from Fort Yuma to resupply them.

On March 9-11, while the main body of the expedition, including Möllhausen, camped at the "head of navigation," Ives, Captain Robinson, and one of the steamboat's crew explored Black Canyon in a rowboat fitted with a blanket for a sail. Passing below the towering canyon walls, the trio reached Las Vegas Wash (which Ives mistook for the Virgin River) beyond the site of present Hoover Dam, then returned downriver to the main camp.

From the rowboat reconnaissance, Ives concluded that the Old Spanish Trail from California to Utah passed not far to the northwest of Black Canyon. However, any further exploration of the river or attempt to reach the Mormon Road would have to be made overland by a different route in order to bypass some of the impenetrable canyon walls. The lieutenant wrote at length about the scenery he had witnessed, likening the "romantic effects" to those seen earlier in Mohave Canyon, "but on an enlarged and grander scale." In a rough cartographic sketch in his topographical notebook, Ives gave the biblical labels "Gog" and "Magog" to "two mammoth peaks" which, he noted in his journal, "seemed to be nodding to each other across the stream"; he gave another formation the name Fortification Rock. In the canyon, he wrote, "[t]he solitude, the stillness, the subdued light, and the vastness of every surrounding object . . . produce an impression of awe that ultimately becomes almost painful." Since neither Möllhausen nor Egloffstein accompanied him in Black Canyon, Ives himself made a rough sketch to convey some idea of it. Egloffstein later redrew it for the final lithograph that appeared in the official report.[144]

On March 13, the entire expedition started downriver on the *Explorer* to meet the pack train, which still had not appeared. Two days later, while they camped in the Cottonwood Valley, Ives' men sighted some tracks belonging to what they believed to be four white men and their animals. Soon afterwards a lone white man mysteriously appeared, whom the explorers recognized as a Mormon agent although he pretended not to be one. Playing along with his ruse, Ives and his men allowed the man to spend the night in their camp before he continued on his way the next day.

The suspicious visit left all involved, including Möllhausen, feeling a lot less secure—and for good reason. The mysterious visitor had in fact been a Mormon named Thales Haskell. Moreover, he had four comrades, including Jacob Hamblin—one of Brigham Young's most trusted scouts—observing the encounter from the safety of a nearby willow thicket on the opposite bank. Brigham Young had sent this small Mormon patrol not only to observe the Ives expedition but also to turn the Mohaves and Paiutes against the American soldiers.[145] In fact, Ives and his men soon witnessed a change of attitude among some of the Mohaves as the *Explorer* continued back downriver.

On March 18, the expedition finally linked up with Peacock's mule train and its escort under Lieutenant Tipton. The weary *arrieros* (muleteers) and soldiers had had a difficult journey through the deserts and mountains along the river's banks and had been harrassed by a small band of Mohaves who attempted to stampede the herd. The combined camp near present-day

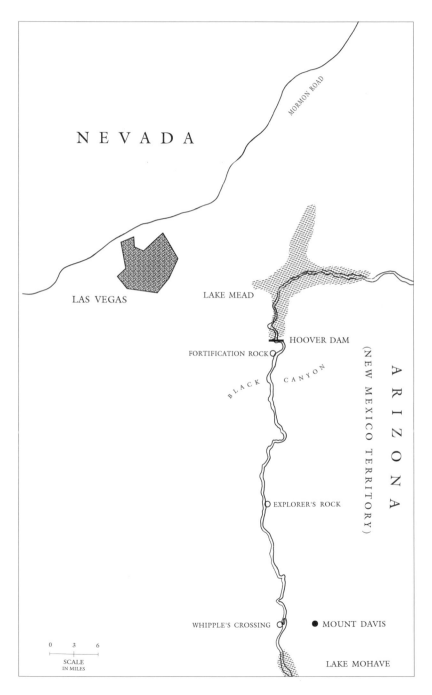

NEVADA

LAS VEGAS

LAKE MEAD

MORMON ROAD

HOOVER DAM

FORTIFICATION ROCK

BLACK CANYON

ARIZONA

(NEW MEXICO TERRITORY)

EXPLORER'S ROCK

WHIPPLE'S CROSSING ● MOUNT DAVIS

LAKE MOHAVE

0   3   6
SCALE
IN MILES

Riviera, Arizona, became a hub of activity for several days as large crowds of Mohaves mingled with them to trade. There was a particularly tense moment on March 22, when some Mohaves with bows and arrows shot and wounded two of the expedition's mules. Fortunately, many on both sides, including Ives, Möllhausen, Cairook, and Iretéba, had a strong desire for peace, and their skillful diplomacy, sensitivity, and watchfulness helped to resolve the problem without any shedding of human blood.

Just how much Möllhausen and the other expedition members actually may have feared for their lives at this time may be inferred from a private letter he wrote his wife in the summer of 1858, at the end of the expedition. The earlier letter to which he refers no longer exists, but one may presume that it terrified his wife:

> You said my last letter was written in a melancholy frame of mind. You were right, my little angel, but I also believed that it would be my last, for we were only 55 men. We camped in woods on the bank of the Colorado and were surrounded by 2000-3000 Mohaves, who were incited by the Mormons, and who owing to their fearful painting and their weapons of war, and the absence of their women and children and owing to the shooting of one of our mules, only too clearly gave evidence of their hostile intentions. We were at our posts night and day. In my game bag lay 30 greased bullets, in my small leather pouch 36 rounds of buckshot, in my vest pocket 50 revolver cartridges, and after I had honed a long knife as sharp as a razor, I sat down with my back against a tree; on my right arm my rifle, on the left a double-barrelled shotgun, on my knees paper, and thus I wrote you, in the face of a terrible, wild, aroused, hostile horde of Indians; in the face, as I then believed, of certain death, for there were certainly too many against us few men. . . . However, everything turned out all right.[146]

To finish his mission, Ives now divided his force. Though well aware of the inherent danger, he sent half of his men, including Bielawski, back downriver with the *Explorer*. The remainder would travel east and northeast overland on mules to determine more about the river and to find another route into the Mormon country. This group included Ives, Möllhausen, Newberry, Egloffstein, Peacock, a Yuma Indian named Yuckeye, two cooks, two servants, eleven Mexican and Californian *arrieros*, and an escort of twenty to twenty-four soldiers commanded by Lieutenant Tipton. In addition, Iretéba and two other Mohaves, Kolhokorao and Hamotamaque, agreed to serve the overland expedition as guides. The train itself consisted of approximately 150 mules.

On March 23, Ives and his men gave a last farewell to Cairook and a large number of Mohaves who turned out to see the explorers off. "We all felt regret at parting with him, for he had proved himself a staunch friend," Ives wrote. This farewell would assume more poignancy in light of Cairook's death the following year, while he was the army's prisoner at Fort Yuma (see cat. no. 24).[147]

With Iretéba and his Indian guides in the lead, Ives' overland expedition headed east and northeast along Beale's wagon road, travelling through "Beale's Pass" in the Black Mountains and through the arid Sacramento Valley to "Railroad Pass"—so named because Whipple had thought it would serve that purpose—in the Cerbat Mountains near the present town of Kingman. From there, Ives' expedition continued in roughly the same direction through Hualapai Valley (Ives' "Cerbat Basin" and Newberry's "Yampai Valley") and followed Beale's trail around the northern end of the Peacock Mountains (named for Ives' mule driver), past or near the future townsites of Hackberry, Valentine, Crozier, Truxton, and Peach Springs (all today along Arizona Highway 66).[148]

The character of the landscape and its inhabitants had completely changed from that of the lower Colorado valley, and the Mohave guides, now far from their own territory, were less and less familiar with the terrain. On March 27 the expedition encountered snow for the first time. Four days later, near Peach Springs, they made contact with some Hualapai Indians. Both Ives and Möllhausen gave these mountain- and canyon-dwelling people unfavorable assessments, partly for their physical appearance and partly because fear, mistrust, and suspicion pervaded all of their dealings with the whites. Ives described them as "squalid, wretched-looking creatures, with splay feet, large joints, and diminutive figures," but "bright and cunning faces." Similarly, Möllhausen described the men as "filthy-looking fellows" who "looked like haggard, prematurely aged children next to our sturdy Mohaves."[149] According to Ives, Möllhausen likened one particularly ugly Hualapai man to a toad and suggested in jest that they "preserve him in alcohol as a zoological specimen."[150] The Mohave guides also referred to the Hualapais with "comical contempt" and advised the whites not to trust them. Not surprisingly, the Hualapais

themselves showed little enthusiasm for the momentous occasion, and Ives had considerable difficulty persuading two of them to lead the expedition through the maze of canyons down to the Colorado River, which he understood was not too far to the north.

The two Hualapais reluctantly guided the expedition to Peach Springs Canyon, which they entered on the morning of April 2 and followed to the canyon of Diamond Creek, which in turn would lead them directly into the Grand Canyon of the Colorado. The explorers camped in a cool, thin, grassy strip in the canyon of Diamond Creek, and from a vantage point on a "projecting wall," Möllhausen sketched the camp and a colossal butte nearby (cat. no. 36). That evening, several men from the expedition, including Möllhausen, walked along the creek, following it to its mouth in the nearby Colorado River, where "intractable torrents of water . . . rebounded from rock to rock," producing "whirlpools and foam as they forced their way into the rocky gorge to the south."[151] With this hike they probably became the first white men to reach the floor of the Grand Canyon.[152]

The expedition members remained camped at Diamond Creek while they explored the area. As the first geologist in the Grand Canyon, Newberry had a tremendous opportunity to observe rock stratifications unlike anything he had ever seen, and he reported:

> The cañon of the Colorado at the mouth of Diamond river is but a portion of the stupendous chasm which its waters have cut in the strata of the table-lands. . . . At this point its walls have an altitude of over 3,000 feet above the Colorado. . . . A few miles further east, where the surface of the table-lands has an altitude of nearly 7,000 feet, the dimensions of the cañon become far more imposing, and its cliffs rise to the height of more than a mile above the river.[153]

Mollhausen wrote that where Diamond Creek joined the Colorado River, "[t]he view left us speechless," but he nonetheless managed to record his impressions through several pages of description in his journal.[154] Climbing over the nearby rocks, he and Egloffstein sketched the location from several different angles (see cat. nos. 37 and 38). Ives praised one of Möllhausen's sketches of the mouth of Diamond Creek for giving "a better idea of it than any description."[155]

Although Lieutenant Ives and Dr. Newberry did not mention it in their official reports, Möllhausen described a small, sad incident involving Grizzly, the pet dog that had followed the expedition

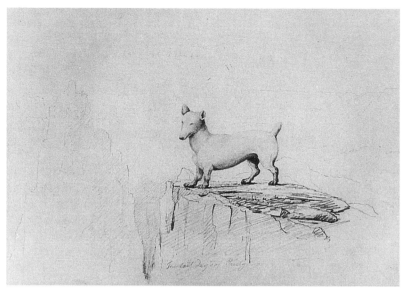

FIG. 58
Balduin Möllhausen, *The Last Days of Grizzly*, graphite on paper, 1858, from
Sketchbook 3, Möllhausen Family, Bleicherode, Germany, photograph
courtesy of Dr. Friedrich Schegk, Munich

from California. Grizzly (fig. 58) was "half dachsund and half wolf-hound [Alsatian] and as hideous an example of his race as I have ever seen," Möllhausen wrote, but over the course of the expedition he had "developed a body like a small calf. Although only a few inches high, his legs were as strong as his powerful body, except that they had a few bends resembling corkscrew."[156] On April 2 Grizzly "was in high spirits" as he chewed on fresh blades of grass near the camp, and early the next morning, while Möllhausen and Dr. Newberry followed the mouth of the creek back to the Colorado, the dog followed Egloffstein as the baron set out with a soldier and an Indian to see the mighty river as well. Egloffstein "selected the most difficult route," trying "to climb up one of the bluffs in order to follow the course of the Colorado somewhat farther and make corrections on his map." By evening, when the men still had not returned, Möllhausen and the others "began to grow concerned. . . . We knew about the experiences he had had on previous expeditions with Colonel Frémont, but we also knew

that he had difficulty bridling his enthusiasm, which often got him into dangerous and unpleasant situations. Climbing down from these steep mountains in the middle of the night was the sort of thing that could get Egloffstein and his companions killed."

Although the men returned to camp late that night "with torn boots, and suffering from sore feet, hunger and thirst," Grizzly had not fared so well. The dog had followed them to

the top of the plateau, but had succumbed to thirst and exhaustion on the way down. Egloffstein and Hamotamaque had carried the poor animal a long way, but when it got dark they were forced to feel their way along the dangerous path, and they had to leave the dog to his fate. Most likely he was either torn to pieces a short distance from camp by hungry wolves or devoured by hungry Walapai Indians, who are equally predatory. Grizzly's loss was painful to us all, for the friendly loyal animal had accompanied us faithfully from the Pueblo of Los Angeles on the Pacific Ocean for more than a thousand miles through a terrible wilderness, had amused us with his affectionate nature and his happy mood. But now we lost him just as he was becoming useful as a watch dog.[157]

Möllhausen's recording of such incidental matters gives his works much of their appeal. Not only did he include a description of the dog and this incident in his narrative, but he even made a field sketch, called *The Last Days of Grizzly* (fig. 58), which rests between other sheets in the sketchbook dated April 5 and 13. Möllhausen probably had sketched Grizzly before his disappearance but added the rocky landscape at a later date.

While the expedition camped at Diamond Creek, more Hualapais came to visit. Ives commented that "their intelligence is of so low an order that it is impossible to glean information from them, and their filthiness makes them objectionable."[158] But Möllhausen perceptively noted that the Hualapai could naturally sense the disdain with which the members of the expedition viewed them, and that they seemed more trusting and sociable with the Mexican *arrieros* and especially with Iretéba.[159]

Once the explorers had seen enough of the canyon to know that they could not cross the Colorado from this location, Ives determined to strike out along the plateau in search of a northeasterly route into the canyonlands. Iretéba persuaded two of the Hualapai to lead the expedition out of the canyon, and on April 4,

Ives and his men followed their Hualapai guides several miles back up the route they had come down, then up a side canyon to the northeast. The following day, Iretéba and his two Mohave companions took leave of the expedition, and the explorers loaded them down with presents during an emotional farewell (see cat. no. 39).

As they crossed the Coconino Plateau (see map below), the expedition encountered changing flora and fauna—including cedars, pines, antelope, and wolves—and heavy spring snows. The Hualapai guides deserted them unexpectedly on the night of April 7, but the next day the explorers found a faint Indian trail leading

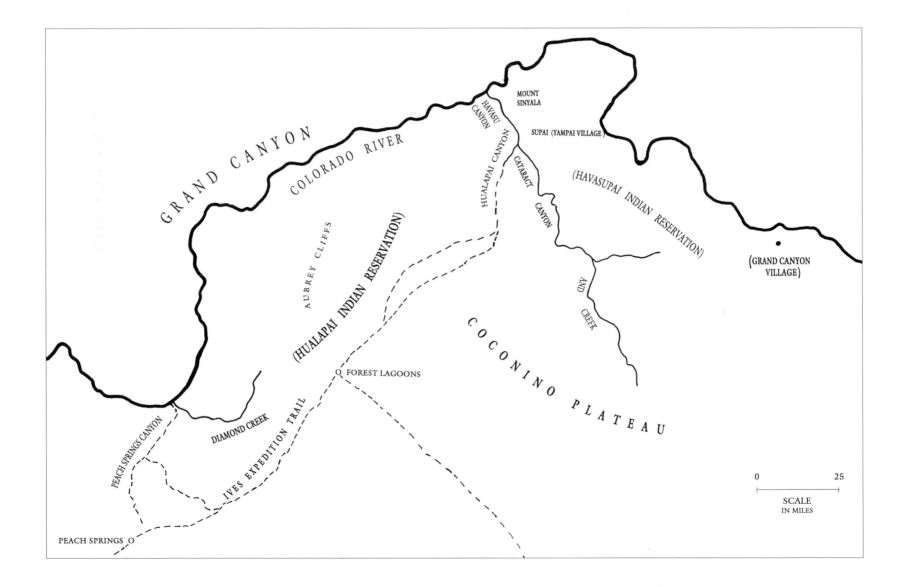

in a northeasterly direction. A snowstorm trapped them in camp on April 9 and 10, but on the move a few days later, they again began to distinguish the deep fissures and vividly colored bluffs of the canyons ahead of them. Ives described the "sublime spectacle" before them:

> Towards the north was the field of plateaus and cañons, already seen and described, and shooting out from these a line of magnificent bluffs, extending eastward an enormous distance, marked the course of the cañon of the Little Colorado. Further south, eighty miles distant, towered the vast pile of the San Francisco mountain, its conical summit covered with snow, and sharply defined against the sky. Several other peaks were visible a little to the right, and half way between us and this cluster of venerable and mightly volcanos was the "Red Butte," described by Lieutenant Whipple, standing in isolated prominence upon the level plain.[160]

On seeing San Francisco and Bill Williams peaks, Möllhausen "stopped and greeted the proud mountains as old friends, and I derived great pleasure from the prospect of being able to hunt on their slopes in the near future."[161] The men also passed by several lagoons, which they would soon discover were the last "accessible watering place west of the mouth of Diamond river." Ives would soon send the mule train back to camp at this spot, and eventually the whole expedition would rest at the "forest lagoons" on the nights of April 16 and 17.[162]

Several miles farther along the plateau, the explorers discovered a mysterious wagon trail crossing their route from a northwesterly or southeasterly direction; they later learned that Lieutenant Beale had blazed it the September before.[163] The Indian trail they were following suddenly led the explorers on a steep descent into Hualapai Canyon, which leads into Cataract and Havasu Canyons and, in turn, into the Grand Canyon. As Hualapai Canyon opened up before them, the trail, precariously situated along one canyon wall, became so narrow and dizzying that the men dismounted to lead their mules in single file. Some of them had vertigo so badly that Möllhausen wrote of "strong men crouching down like faint-hearted children in order to ward off dizzy spells." Ives believed that with such a well-beaten Indian trail indicated "every prospect . . . that we were approaching a settlement similar to that of the Hualpais, on Diamond River," so he pushed on until they reached a point of descent that "no mule could accomplish."[164] With tired and thirsty mules, no settlement in sight, and the nearest known water source nearly thirty miles behind them, Ives decided to turn back.

Fortunately, they soon discovered a small rock cistern of water, which sufficed for the men and a few animals. Ives sent the rest of the mules back to the lagoons with the *arrieros*, then assembled a small party of men, including himself, Lieutenant Tipton, Egloffstein, and Peacock, to reconnoiter the bottom of the Grand Canyon on foot.

Möllhausen and Newberry split off from the group to try on their own to reach the bottom of the canyons. Giving up, they instead climbed the heights near the modern-day Hualapai Hilltop turnaround on the Havasupai Indian Reservation. There Möllhausen sketched a spectacular view of a rocky amphitheater, formed where the Grand Canyon is joined by some of its tributary canyons, and the distinctive butte known today as Mount Sinyala (see cat. no. 42). With his usual thoroughness, he also left a lengthy written description of the entire panorama.[165]

Meanwhile, Ives' reconnaissance party, far below, discovered that the bottom of Hualapai Canyon contained an inner canyon. Still following an Indian trail, they descended, in Ives' words, "deeper in the bowels of the earth than we had ever been before" and were "surrounded by walls and towers of such imposing dimensions" that "it would be useless to attempt to describe them." After thirteen miles the trail led them along a creek to a precipice with a forty-foot drop, which appeared at first to block their further descent. However, as Egloffstein lay down to look over the edge and watch the cascading waters of the creek, he discovered a "crazy looking ladder, made of rough sticks bound together with thongs of bark."[166] Although it was partly rotted, the inveterate Bavarian rock climber was so eager to get the first view from the bottom that he attempted to lower himself down on it. One side of the ladder slowly gave way under his weight and fell crashing to the bottom,

but he managed to lower himself the rest of the way without injury. While Ives and his men devised a way to hoist him back out, the ever-curious Egloffstein scouted out the area on his own.

At the canyon bottom, Egloffstein found a stream with a narrow belt of land, including cottonwoods, willows, irrigated cornfields, and a few scattered huts belonging to the "Yampais" (Havasupai). He had reached Supai village (still very much in existence today, on the Havasupai Indian Reservation). From the description Egloffstein gave later, Ives wrote that the residents, who he estimated did not number over two hundred, were similar to the Hualapai in appearance but "perhaps a trifle cleaner and more respectable."[167] Accompanied by one of the Havasupai, Egloffstein obtained a view of the valley from a ledge, and saw where Cataract Creek flows into the Colorado River. After this, the two men made their way back to the broken ladder at the foot of the precipice where Ives' group had remained. With a rope improvised by tying together the soldiers' musket slings, Ives and his men hoisted Egloffstein out of the bottom. However, they were unable to convince the helpful Havasupai man to accompany them.

After spending the night in the canyon, Ives and his group returned on April 14 to the main camp, where Möllhausen and Newberry awaited them. Möllhausen, surmising that the village was near the red sandstone expanse beneath a rocky butte shown in *Doctor's Rest* (cat. no. 42), found that his sketch "aroused much interest" when he showed it to Ives and his party.[168] That same afternoon, Möllhausen, Newberry, and Egloffstein (despite the latter's sore feet) headed off in an easterly direction in hopes of obtaining another view of the Colorado. This time, after about a three-mile march, they reached the head of Driftwood Canyon, where Möllhausen made a sketch (*Cañon near Upper Cataract Creek*), looking northeast and showing the actual Grand Canyon in the distance (cat. no. 43).

Having determined that rugged canyons blocked their way north and east into Mormon country, the expedition returned to the "forest lagoons" where the mule train was resting and spent the next three days camped there. On the morning of April 19, they headed south along the cold, dry plateau. Lieutenant Ives had decided to visit the Moqui (Hopi) Indians, and he believed that the best way there would be to cross the Little Colorado east of San Francisco Mountain, about ninety miles to the southeast. Passing through an "eroded plateau, marking the northern boundary of the volcanic belt of the San Francisco and Bill Williams mountains," Möllhausen commented on the "strangely beautiful scenery" of the distant snow-covered mountains and the "many small volcanic cones scattered around them."[169]

They soon discovered and followed a faint trail blazed six months earlier by Beale and his camels. This led the explorers southeast in the direction of the snowy peaks of San Francisco and Bill Williams' Mountains, visible in the distance and already well known to those who had been on Whipple's expedition. In fact, shortly after April 21, when they reached Partridge Creek, Ives and his men began generally to follow Whipple's route east toward the Little Colorado or Flax River.[170]

On the morning of April 21, Möllhausen, Peacock, and Dr. Newberry were riding ahead of the main party in a wide canyon when they discovered fresh bear tracks. According to Möllhausen, they had just begun to track the animal when they "suddenly caught sight of the bear itself carelessly plucking sweet plants out of the ground in a bend formed by a side canyon. He was a huge fellow, and we could clearly follow his movements because his broad black back poked out above the dry underbrush."[171] The three men checked their weapons and surrounded the animal. Unknown to them, however, half of the company of soldiers behind them had also seen the bear and rushed after him. "The grim monster" was "quietly ascending a hill," Ives recounted, when "a volley of balls made the fur fly in all directions from different parts of his hide. Twice he turned as though meaning to show fight, but the crowd of pursuers was so large, and the firing so hot, that he continued his flight to the top of the hill, where he fell dead, riddled with bullets."[172] Möllhausen was "quite annoyed that a nice hunt . . . had been more or less spoiled."[173]

While some of the men skinned the bear and divided the meat, others set up camp in a side ravine where "a clear spring bubbled up from the volcanic rock." Möllhausen and Ives believed the main canyon was the dry bed of Partridge Creek—a stream they remembered from the Whipple expedition in January 1854.[174] They spent most of the rest of the day in camp and dined on bear meat, which Ives remarked was "rather too strong flavored to be palatable when roasted or broiled, but makes capital soup"—an observation Möllhausen later confirmed when he recalled "a heavily seasoned but nourishing soup from the bones of the bear

FIG. 59
Balduin Möllhausen, *Grizzly Bear on Partridge Creek [Grizzly am Cardridge Creek]*, graphite on paper, 1858, from Sketchbook 3, Möllhausen Family, Bleicherode, Germany, photograph courtesy of Dr. Friedrich Schegk

FIG. 60
Balduin Möllhausen, *Grizzly Bear on Partridge Creek [Grizzly am Cardridge Creek]*, graphite on paper, 1858, from Sketchbook 3, Möllhausen Family, Bleicherode, Germany, photograph courtesy of Dr. Friedrich Schegk

from Partridge Creek. In many ways it reminded us of the turtle soups of New Orleans."[175]

Möllhausen made two sketches of the bear on Partridge Creek, probably from observations of the carcass (figs. 59 and 60). These drawings may have also benefitted from his witnessing a second bear hunt on April 26, between Bill Williams Mountain and the San Francisco Mountains, recorded in a watercolor (see p. 132). On this occasion, Möllhausen recorded that a Mexican *arriero* named Fernando killed a full-grown bear, only to have the soldiers again riddle the dying animal with bullets and then claim the honor of having made the first shot![176] From his sketches, Möllhausen finished a watercolor in 1859; it hung framed in his study for several years (see fig. 71).[177]

Ives also noted the presence of antelope tracks in his description of Partridge Creek, and Möllhausen recorded on April 22 that as they travelled along a dry ravine or canyon, "[b]eautifully marked antelope ran from the canyon on both sides as we approached, and . . . looked down on us with curiosity from a safe distance."[178] Möllhausen seems to have sketched a pronghorn antelope on April 24 (fig. 61),[179] and a day or two later, he drew a view of Bill Williams' Mountain and the expedition's snowy camp amid the pines (cat. no. 44).

The explorers arrived at the Little Colorado River near the present-day town of Leupp on May 1. Here the expedition separated again, partly because supplies were running low. Ives took a small reconnaissance party—Newberry, Egloffstein, a few soldiers, *arrieros*, and the best animals—across the river and turned northwards towards the Moqui (Hopi) villages, on a long and circuitous route to Fort Defiance, the expedition's official destination. The rest of the expedition, including Möllhausen, Peacock, and Lieutenant Tipton, continued east along Whipple's route toward Zuñi pueblo. From there, they would go north to rendezvous with Ives at Fort Defiance.

By the middle of the nineteenty century, the Hopi pueblos were still among the lesser known and more isolated Indian villages in the American Southwest; probably for this reason and because he had never been there, Möllhausen "felt a greater inclination to visit the Moqui." However, since the artist was the only expedition member besides Ives who already knew the route to Zuñi, Ives preferred that Möllhausen guide the other group. This was somewhat unfortunate from the standpoint of new visual documentation,

however, because Ives' contingent would have only the graphic services of Egloffstein and Newberry when they visited this little-known area (figs. 62 and 63),[180] while eyewitness images of Zuñi pueblo already existed, including some by Möllhausen (fig. 64).[181]

Travelling north and east across the Painted Desert, Ives' party evidently missed Möllhausen's presence, not just for his sketching ability but also for his specimen collecting. In the cooler temperatures of the night, Ives noted that the normally hot and empty desert came alive with scorpions, spiders, rattlesnakes, and centipedes, and a "collector in that department of natural history could have reaped a harvest of these reptiles in almost any part of our camp-ground."[182] Since they travelled so rapidly, without Möllhausen to assist, Newberry probably had to run faster than usual to collect geological specimens.[183]

On May 11 Ives and his men camped near two of the seven Hopi pueblos. The lieutenant, Egloffstein, and Newberry entered "Mooshahneh" (Mishongnovi) pueblo at the invitation of the friendly inhabitants. Ives was intrigued by the neat appearance of the people, with their colorful clothes, flocks of sheep, peach trees

Fig. 61
Balduin Möllhausen, *Pronghorn Antelope [Antelope]*, graphite on paper, from Sketchbook 3, Möllhausen Family, Bleicherode, Germany, photograph courtesy of Dr. Friedrich Schegk, Munich

and garden terraces, and unusual home architecture. In addition to his usual landscape panoramas, Egloffstein made a rather detailed sketch of the interior of one apartment they visited. Ives and his group spent several days among the Hopi, but he was unable to find someone willing to guide his group to the Colorado River. After visiting the pueblos of Oraibi and Tewa, the group headed east towards Fort Defiance, accompanied by an ever-increasing number of Hopis and Navahos.

Meanwhile, the southern contingent traversed the Petrified Forest and camped near the pueblo of Zuñi. They encountered several old acquaintances along the way: Whipple's former guide, José Manuel Savedra, who was heading west with two companions in search of three or four mules who had escaped from Beale's expedition; Zuñi chief José Maria; and Pedro Pino (fig. 66), the Indian governor of Zuñi pueblo.[184] On May 11, Möllhausen visited the interiors of some buildings at Zuñi, an opportunity he had not had earlier with Whipple because a smallpox epidemic was raging there. The scene reminded him of "a beehive's interior, in which the industrious insects scurry back and forth from cell to cell." The cleanliness and simplicity of the rooms and the generous hospitality of the people highly impressed him. He was embarrassed at his own shabby appearance and the complete lack of provisions, which prevented him from reciprocating the favors they bestowed upon him. Möllhausen was especially struck by their apparent happiness and their ability to preserve their ancient traditions:

> "Happy people!" I thought again. "Happy people in your semi-civilized condition. May European civilization remain far from your peaceful homes."[185]

This simple hope was not entirely possible, as Möllhausen already knew. European contact had already had an effect, for example, on the appearance of Chief José Hatche, another Whipple guide and friend of Möllhausen's, whom the German hardly recognized because smallpox (of European origin) had badly disfigured his face in the intervening years. As further proof of changing times, Möllhausen recorded that while at Zuñi, one of the expedition's *arrieros* was murdered by another compadre after both had been drinking.[186]

Accompanied by a Navaho guide, Möllhausen's group left Zuñi on May 12 and two days later reached Fort Defiance, where they waited until May 22 for Ives' contingent. Möllhausen continued

FIG. 62
John James Young after F. W. von Egloffstein, *Moquis Pueblos*,
transfer lithograph, from Ives, *Report* (1861), part 1, opp. p. 119,
Amon Carter Museum Library

FIG. 63
John James Young after F. W. von Egloffstein, *Interior of a Moquis House*,
toned lithograph (black with red and gold), from Ives, *Report* (1861),
part 1, opp. p. 121, Amon Carter Museum Library

FIG. 64
After H. B. Möllhausen, *Zuni*, lithograph, in Whipple report, *Reports of Explorations and Surveys* (Washington, 1856), vol. 3, part 1, opp. p. 67, Amon Carter Museum

FIG. 65
Balduin Möllhausen, *Natural Bridge, 17 mai*, graphite on paper, from Sketchbook 3, Möllhausen Family, Bleicherode, Germany, photograph courtesy of Dr. Friedrich Schegk, Munich

sketching: among his subjects were picturesque rock formations, the fort, some Hopis and Navahos he encountered,[186] and scenes showing one of the explorers getting a well-deserved haircut and two of them loading a mule (figs. 65-69). The two groups had travelled for miles through intermittent Navaho country without incident, and only after their safe arrival and equally pleasant departure from Fort Defiance did it become clear how fortunate this was. Five days after they left the fort on May 24, a dispute over grazing lands between the fort's commander and the Navaho headman erupted into outright hostilities.[188]

Although the expedition was officially over, most of its members still faced the difficult problem of getting back to Washington, D.C., with their collections, notes, and sketches, and Ives still had to pay his men and take care of further business in California. The expedition headed for Albuquerque[189] and camped there for the first ten days of June, while Ives went to Santa Fe to withdraw funds. During this time, Möllhausen, who had been in Albuquerque with Whipple in 1853, attended fandangos, participated in the Corpus Christi celebrations, and described and sketched some of the inhabitants. Ives soon returned and took leave of his men on June 8 to catch a stage coach to California via El Paso. Stopping at Fort Yuma to sell the *Explorer* to Johnson for $1,000, the lieutenant continued on to San Francisco to begin the long steamer trip back east.

The civilian members of the expedition had no reason to go back west with Ives, so they had decided to follow the Santa Fe Trail across the Great Plains. This party—Möllhausen, Egloffstein, Newberry, Peacock, two Irishmen named Wigham and O'Connor who had accompanied them from San Francisco as cook and servant, and a young American roustabout named Hendricks, who was employed as teamster—left Albuquerque on June 11 with six riding mules and a wagon drawn by eight mules. At Santa Fe they rendezvoused with Lieutenant Tipton and a few soldiers, and at Fort Union they were joined by an officer and his family who were heading east.

This trip across the Great Plains, although dangerous, was less eventful than Möllhausen's earlier journey with Duke Paul. The artist enjoyed his last buffalo hunts, swapped campfire stories with the others, and visited with westward-bound emigrants, traders, and soldiers. Near the Arkansas River they saw some Cheyennes, Arapahos, and Kiowas. Möllhausen conversed with one of the

Fig. 66
Balduin Möllhausen, *Governor Pedro Pino of Zuni on a Donkey [Berittener Krieger]*, graphite on paper, from Sketchbook 3, Möllhausen Family, Bleicherode, Germany, photograph courtesy of Dr. Friedrich Schegk

Fig. 67
John James Young after Balduin Möllhausen, *Moquis*, chromolithograph (black with blue, red, and gold), from Ives, *Report* (1861), part 1, opp. p. 120, Amon Carter Museum Library

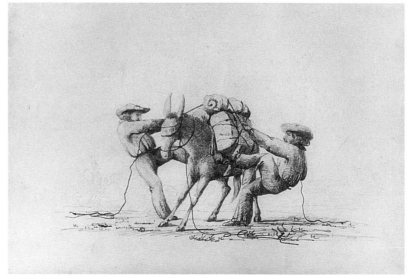

Figs. 68-69
Balduin Möllhausen, activities at Fort Defiance, graphite on paper, from Sketchbook 3, Möllhausen Family, Bleicherode, Germany, photograph courtesy of Dr. Friedrich Schegk, Munich

Cheyennes and sketched his portrait, which along with the studies of buffalo in his sketchbook are probably the last images he drew in America.

Passing through the Kansas Territory, which was already embroiled in the controversy over slavery, Möllhausen opined:

> Whether the free man eats his bread by the sweat of his brow, or whether the colored slave winces under the whip of a cruel master, no barrier can stem the tide of civilization any more than it can stem the final solution of the slave question. The solution of this problem may be artificially held in abeyance during the coming decades, but slavery as an unnatural institution must break down of its own accord, especially in this time of progress and of growing enlightenment.[190]

The group reached Fort Leavenworth on July 24 and disbanded shortly thereafter. Möllhausen and Newberry travelled by steamer down the Missouri to St. Louis, then by train to Cleveland, Ohio, where Möllhausen spent eight days with Newberry's family before taking the train to New York City and Washington, D.C. The expedition itself was over, although much work remained to be done.

FROM SKETCHES TO PRINTS

By the end of the expedition, Möllhausen was in great haste to return to his homeland. The long separation from his wife and child had been almost unbearable, as he wrote his friend Spencer Baird on August 11 from New York City:

> I am very unfortunate and really heartbroken, all my letters have gone to Fort Yuma, to what place we first intended to return, and now I am here with the latest news from the 27th of November, is it not horrible? Baron von Gerolt told me that my family is well, but that don't make me much easier. I have no quiet hour neither day or night till I hold in my hands the proofs that my wife and boy are well. A married man ought not to travel, it is hard to be away so very far from wife and child.[191]

Möllhausen also was eager to see and share his experiences with his friends and benefactors in Prussia. While at Fort Yuma he had

received the sad news of the death of his old friend and mentor Lichtenstein. Two of his other benefactors were in poor health: Humboldt was now almost ninety years old, and King Friedrich Wilhelm IV had had a mental collapse after a stroke.[192] For these reasons and probably others, Möllhausen sought permission from his American employers to complete his artwork for the expedition back in Prussia rather than in Washington, D.C.

This permission became a matter of great concern. On July 20, 1858, even before Möllhausen and some of the expedition arrived in Washington, Ambassador Gerolt had written Humboldt:

> Lt. Ives writes that Möllhausen brings with him a rich collection of views and natural history specimens. Whether he will be able to work up these things in Berlin—as Mr. Floyd lets me hope, appears to me nevertheless somewhat doubtful. This is on account of the declarations of the lower officials who fear that misuse will be made of these things. Lt. Ives' opinion over this will probably determine the Secretary of War's decision and some lines by your Excellency regarding this would probably turn the scale.[193]

From Washington on August 20, Möllhausen wrote his wife:

> Your letter to Miss Corcoran has had an effect. Today I see the Secretary of War, and if envious people do not work against me, I will, according to the promise of Mr. Floyd, be able to finish the works in Berlin. If not, on the other hand, I shall refuse. Ives has still not arrived, and I will secure myself money through Corcoran and settle matters with Ives by letter. I *must*, I *must*, I *must* travel on September 1st.[194]

Part of Möllhausen's difficulty in obtaining permission was that he had to await Ives' arrival in New York City. Toward the last of August the two men finally met there at a prearranged hotel and agreed on a date for Möllhausen to deliver the completed works.[195] These special arrangements and Möllhausen's divided loyalties may account for some of the apparent personal tension between him and Lieutenant Ives toward the end of the expedition. Möllhausen criticized some of Ives' decisions in *Reisen*, and he may well have expressed these criticisms to his American commander. His outspoken antislavery views also must have conflicted with Ives' prosouthern sympathies, and no doubt fatigue and many weeks of constant contact also contributed to this friction.

On August 29, 1858, Möllhausen wrote Spencer Baird:

Mr. Ives has come at last, and we have arranged my affairs in the most pleasant way; on the first of September I leave in the Steamer Saxonia for Hamburg taking along only agre[e]able recollections, and forgetting small disagreeable circumstances, which always accompany expeditions t[h]rough the vast western regions. I do not take my sketches with me, but it is more than likely, that Mr. Ives will forward them within a short time through the Prussian Legation.[196]

Möllhausen had from five to six months to work on the watercolors in Prussia, and true to his word, by March of 1859 he had sent the completed works to Ives.[197]

Möllhausen's absence from Washington during the final work on the Ives project had both benefits and drawbacks. He was near his family and probably enjoyed painting in familiar surroundings. He also could copy or make slightly different versions of the watercolors for his own personal use, and during this time he prepared a set of sketches to illustrate *Reisen*.[198] Although Möllhausen could seek more artistic advice from Hildebrandt and others in Berlin and Potsdam, his absence from Washington obviously prevented him from consulting any of his fellow expedition members for details or opinions about his sketches. Moreover, since his field sketches contain no color notes, he must have been hard-pressed to remember colors correctly. He could not add new sketches or make any alterations once he sent the watercolors to the United States.

Considering the time and distance between the creation of the field sketches and execution of the watercolors, it is surprising that the watercolors retain any fidelity to nature at all, yet they do. Viewed in conjunction with the prints, the written descriptions by Ives and Möllhausen, and nineteenth- and twentieth-century topographical maps, many of the sites in the watercolors are still identifiable. Photographs taken in 1990 demonstrate some topographical and geological inaccuracies but also confirm the considerable realism of the painted views. Moreover, the watercolors sometimes bring the narratives to life, providing more information than any of the written accounts, while at other times the reverse is true.

While the catalogue analyzes individually and in depth all forty-six of the Amon Carter Museum's watercolors, a few generaliza-

tions about these images seem relevant here. Möllhausen had developed some stylistic techniques when creating works on his second trip to America in 1853-54, but his Ives expedition watercolors, like the ones now in Potsdam and Berlin, show noticeable improvements over his earlier efforts. For example, in *Steamboat Explorer (Chimney Peak)* (cat. no. 6), the various ellipses on the steamboat, while not perfectly executed, nevertheless demonstrate a fair knowledge of perspective when compared to Möllhausen's 1851 sketch of a steamboat on the Mississippi (see fig. 4). The watercolor showing Ives' little steamboat puffing its way past Chimney Peak combines two artistic genres, the ship portrait and the topographical landscape, and probably was composed from at least two separate sketches (both presumably lost), one of the steamboat and another of the landscape. The human figures on board the later steamboat are either relaxed or animated, depending upon the nature of their activity, and for their size, are all reasonably well proportioned. Many details, such as the man at the bow with the sounding pole, the fireman stoking the boiler, and the pilot leaning upon the tiller, also turn up in Möllhausen's and Ives' lengthy written descriptions of the boat and this well-known topographical landmark.

The rather bland coloring of this image is typical of the watercolors that Möllhausen created back in Prussia. That he usually worked without color notes did not matter very much for his purposes: the printmakers generally did not need to know the exact coloration of objects because the completed prints themselves contained three or four colors at most (and as it happened, most of his images were made into monochrome wood engravings or monotone or duotone lithographs). To tone his graphite underdrawings, Möllhausen relied on only a few colors—burnt umber, yellowish-brown, bluish-green (occasionally combined with the yellowish-brown to create an olive green), opaque white, and pinkish-orange (probably made by diluting the light cadmium red that occasionally appears in details such as the steamboat's paddle wheel).

In his riparian landscapes, such as *Steamboat Explorer (Chimney Peak)* and *Purple Hill Pass, View Towards North* (cat. no. 3), he laid in some of the areas in the water and skies with transparent washes, at times allowing the slightly raised texture of the tan, medium-weight "De Canson" watercolor paper to show through or create highlights. In still other areas, such as the ripples in the water, the frothy wake behind the churning paddle wheel, or the billowy clouds, he used pure opaque white. He also combined opaque white with other colors to highlight figures or objects. Occasionally he used a dry-brush technique that combined with the slight texture of the paper to create a rough, textured effect.

Möllhausen often achieved a competent aerial perspective in his works by contrasting dark foregrounds with lighter distant landscapes. For example, the distant riverbank and mountains in *Steamboat Explorer (Chimney Peak)* recede convincingly behind the boat, while in the foreground, dark, burnt umber rocks and a fallen tree trunk serve as foils or framing elements (répoussoir) for the composition. In *Purple Hill Pass, View Towards North*, dark olive and umber cliffs in shadow and a dark, protruding finger of riverbank with vegetation lead the eye upriver into the fading distant landscape. The principal colors in the picture include umber brown, bluish-olive green, pinkish-orange, opaque white, and the tan of the paper.

The subdued colors in Möllhausen's sketches are disconcerting, considering that he and the rest of the explorers wrote of the great variety of colors exhibited in the southwestern landscape. Passing the Riverside Mountains (cat. no. 13), Möllhausen wrote of "garish colors . . . especially the play of the violet, blue and red" and of "stark barren slopes [that] glittered from every vantage point."[199] Unfortunately, his watercolor of this scene relies on his usual browns and bluish-olive green and conveys little of this colorful variety.

Möllhausen's few night scenes contain stronger contrasts and a more interesting use of color. Beginning with deep blue-greenish-black landforms and shadows, the artist added opaque white, reddish-orange and yellow. In addition, he would often scrape the paper to further denote campfires and highlights. His night scenes suggest a familiarity with outstanding examples in European art. His moonlight composition *Chimney Rock, From the North* (cat. no. 8), if reversed, would strongly resemble one of the influential prototypes of this genre—Adam Elsheimer's *Moonlight Landscape with the Flight into Egypt* (1609), which Möllhausen could have seen in the Alte Pinakothek in Munich. However, he need not have seen the original or even a print of it to have been strongly influenced by similar works by later romantic German painters.

The fine watercolor titled *Character of the High Table Lands (Camp—Colorado Plateau)* (cat. no. 40) exemplifies Möllhausen's ability to paint a wintry snow scene. A comparison of this watercolor with his original field sketch, *Landscape on the High Plateau, April 10th* (see cat. 40a), gives an idea of his working methods. In the field sketch, he recorded the essential landscape components; for the finished watercolor, probably working from other sketches or written notes, he added figures, campfires, a pyramidal tent, and some dead trees. Möllhausen skillfully used diagonal strokes to indicate evergreen foliage in the graphite sketch and blocked in dark greens and umbers to denote the same in the watercolor.

In several of the Ives expedition watercolors, Möllhausen demonstrated a stylistic technique for rendering reflections that he apparently had not employed before. This consisted of abrupt, vertical straight edges dividing dark and light reflections in the water. As is typical of artists who have learned a new technique, he used it repeatedly. Where exactly the artist learned this device is not certain, but it is not unusual. An aquatint in Humboldt's atlas employs it in depicting a raft on Ecuador's Guayaquil (Guayas?) River. However, Möllhausen could just as well have learned it from Hildebrandt or a watercolor manual.

Although Möllhausen was often accurate in his rendering of distant topographical outlines, the lower landforms in his pictures are more difficult to identify and compare to the actual site. This is partly because some of these areas are less distinctive than the major landmarks and thus harder to locate, and partly because changing vegetation, eroding floodwaters, and dams built in the intervening years have altered much of the landscape seen in Möllhausen's foregrounds.

Despite their sometimes remarkable verisimilitude, his hand-drawn images also present a problem of scale if taken literally. Invariably, the artist brought distant landforms into closer view than they actually appear to the eye—an artistic device known as telescoping, which is also characteristic of a natural historian, who collapses intervening space to examine structure and to focus attention to fragments, details, and particulars.[200] Thus Chimney Peak (Picacho), the Castle Dome Mountains, the Needles, and other landmarks look larger in the watercolors than in photographs. On the other hand, it is also important to note that standard camera lenses tend to flatten objects and make them appear more distant than they really are. Thus, Möllhausen's images are sometimes not as far from reality as they may appear when compared to photographs.

On the rare occasions when the *Explorer* could get up a full head of steam and advance rapidly up the river, Möllhausen was apparently hard-pressed to complete a sketch. This evidently caused some sacrifice of accuracy in his works. A field sketch inscribed simply "3te Canon, Mohave vall" (cat. 23b) is little more than an outline of the geological formations, probably looking northwest up Mohave Canyon near Needles, California. The steamboat quickly passed the section of river depicted in *Left Bank, Four Miles South of Bill Williams Fork* (cat. no. 18)as well as the formations anthropomorphized in *Sleeper's Bend* (cat. no. 10). Unlike a trained geologist, Möllhausen generally concentrated on overall topographical outlines rather than rock textures or specific geological features.

Despite his occasional minor departures from strict fidelity to nature, Möllhausen's watercolors effectively chronicle some of the environmental changes that time has wrought. A number of the river sites are partially underwater today: the Boat Rock formation in the foreground of *Mount Whipple and Boat Rock* (cat. no. 20) is now beneath Lake Havasu, and most of the land appearing in *Cottonwood Valley* (cat. no. 28) is now on the bottom of Lake Mohave. The two views of Monument Canyon (*Left Bank, Four Miles South of Bill Williams Fork* [cat. no. 18] and *Four Miles North of Bill Williams Fork, View Down the River* [cat. no. 19]) show saguaros along this stretch of the river, but such cacti are either nonexistent or extremely rare at these sites today.

Considered together, Möllhausen's pictures and the descriptive accounts of the Ives expedition relate other interesting aspects of southwestern natural history. Although he no doubt added the bighorn sheep in the lower right of *Lighthouse Rock* for pictorial effect, the expedition members actually did encounter bighorns near there. (Karl Bodmer's earlier image of bighorns among the cliffs along the Upper Missouri River also could have partly inspired Möllhausen's picture). While Möllhausen often used the formulaic device of two tiny brushstroke Vs to indicate birds in flight (see, for example, *Purple Hill Pass, View Towards North* [cat. no. 3]), he occasionally showed identifiable birds. Thus, several aquatic birds—probably cranes[201]—take flight in the center foreground of *Riverside Mountain* (cat. no. 13); white egrets or herons, with their long beaks and familiar, S-shaped long necks, appear in the foreground waters of his views of the Monument and Mohave mountains (cat. nos. 15 and 21); and a bald eagle soars above the canyon depths in the watercolor titled *Doctor's Rest* (cat. no. 42).

Occasionally the watercolors depict a specific object of nature other than the landscape itself, such as the two views of large, single trees (cat. nos. 45 and 46). Other Möllhausen paintings have animals as their focal points, such as his watercolors of grizzly bears, pronghorn antelopes, or wandering buffaloes.[202] In general, as he demonstrated in these late examples, he was quite skillful in rendering natural subjects realistically.

Möllhausen included interesting ethnographic details in several of his landscapes, and he nearly always described these details in his journal. A raft with three Mohaves appears in the watercolor *Red Rock Gate* (cat. no. 9); another raft with a Mohave family floats by in the watercolor *Twelve Miles South [sic., North] of Round Island (Gravel Bluffs South of Black Mountains)* (cat. no. 31). A group of Mohaves go for a swim in *The Needles, View from the North* (cat. no. 24), and several demonstrate their fishing technique in Möllhausen's watercolor of *Cottonwood Valley*.

More important ethnographically and historically is *Mohave Indians* (cat. no. 12), one of Möllhausen's most colorful compositions. The artist daubed the brown flesh of the natives with dark blue stripes and cadmium red, and used such colors as olive green in the trees, dark blue and light blue in the clouds and water, yellow green on the riverbank, and touches of opaque white. This watercolor offers important similarities to two Möllhausen watercolors of Mohave Indians among the six in the Berlin Museum of Ethnology.[203] The physiognomies and ethnographic details are similar, as are many of the colors employed, and the warriors in all three watercolors may have been partly inspired by an earlier rendering of Mohave warriors by Richard Kern (see fig. 33), who had observed them with Lorenzo Sitgreaves' expedition in 1851. One of the Berlin paintings (see p. 130) depicts a Mohave dwelling with Mohave figures carrying water or playing a ring and pole game; it probably served as a model for a print in *Diary of a Journey* and therefore dates from between 1854 and early 1857.[204] Two other Berlin compositions (see pp. 130-131) could be termed "Indian costume pictures"—a somewhat artificial genre at which Möllhausen became quite proficient. Möllhausen may have originally intended *Mohave-Indianer* (p. 130) as an illustration for *Reisen;* it seems to date stylistically from approximately the same period as the Amon Carter watercolor. His *Hualpai Indians on Diamond Creek* (see p. 131) probably also combined several sketches of figures he had not seen together at one time. Other, even more composite Indian costume pictures pertaining to the Ives expedition were among the Möllhausen watercolors destroyed in World War II and now known only from transparencies.

Many of Möllhausen's techniques and stylistic devices for painting landscapes of the river journey are also found in his watercolors for the overland expedition, despite the differences in topography. In the two watercolors depicting the Colorado River at the mouth of Diamond Creek (cat. nos. 37 and 38), for example, Möllhausen applied his usual umbers, tans, olive greens, opaque blue-grays, and pinkish-orange colors in both the opaque gouache and the transparent wash methods. Both watercolors utilize dark, rocky foreground projections and figures as répoussoir elements to frame the images and provide a sense of scale. Möllhausen effectively rendered rapids in these views with opaque white, occasionally applied in a dry-brush manner. In several areas he sponged away unwanted colors, or he rubbed them—a technique known as "scumbling." In the view taken from the north, Möllhausen painted one of his familiar, straight-edged perpendicular reflections in the water.

The canyon country overwhelmed Möllhausen's modest artistic abilities. On March 12, near the point they designated "the Head of Navigation," the artist despaired of attaining geological or photographic perfection in his landscapes. "I did not," he confessed, "dare try to portray with my pencil the countless lines that faded into one another before my eyes, but which were so clearly delineated. It is possible to reproduce the true character of such a landscape only by using a device in which the reflected picture is captured on a prepared surface so that every line, every elevation and every minute undulation is exactly defined."[205] Nonetheless, he made a valiant and important first effort to convey the magnificence of the Grand Canyon.

Once he completed and delivered his watercolors to Washington, their reproduction and printing involved many individuals in several government offices and at least one commercial firm. As with the Whipple report, Möllhausen's involvement probably stopped once he sent his artwork to his expedition commander. Approval of the prints made from his sketches was completely beyond his control.

Surviving correspondence in the National Archives gives some idea—although not a complete one—of the report-printing process. First, of course, Congress had to appropriate sufficient funds for the undertaking, which could be a problem. In a letter from Santa Fe dated June 2, 1858, Lieutenant Ives wrote of his fear that "Congress has determined not to print any more reports, particularly such as have pictures in them,"[206] and rising reproduction and printing costs increased the possibility that Ives' *Report* might not be illustrated at all. Once the funds were appropriated, the War Department's Office of Explorations & Surveys had control over the entire project. Although Lieutenant Ives had led the expedition in the field, his immediate superior, Captain Andrew A. Humphreys, was in charge of the project and provided funds for Ives to pay members of the expedition. On December 15, 1858, Ives requested an additional $2500 to pay the "geologist" (no doubt Newberry) for four months at $150 a month, the "topographer & engraver" (probably Egloffstein) for six months at $150, a "copyist" (probably John James Young) $100 per month for four months, $400 for "lettering & toning 2 plates for maps," and $200 for "preparing drawings to illustrate Natural History reports."[207]

Ives apparently took Möllhausen's completed watercolors to the Washington office, where he and other expedition specialists like Newberry may have instructed the copy artist to redraw certain works. The sketches also may have been reviewed by experts from the Smithsonian, although little documentation has been found to prove this. After the Office of Explorations & Surveys approved the artwork, it apparently was sent to the Superintendent of Public Printing. The woodcuts may have been done by government engravers, but the lithographs were by the prominent New York firm of Sarony, Major & Knapp.[208] Proof prints were returned, via the Superintendent of Printing, to the Office of Explorations and Surveys for final approval before printing.[209]

In all, thirty-four of the forty-six Amon Carter Museum watercolors served as models for prints in Ives' *Report*. Allowing for the unavoidable differences between the watercolor medium and engraving and lithography, and for the intervention of at least one other human being in the printmaking process, many of the original watercolors were reproduced in Ives' *Report* with seeming fidelity. The engraver naturally had difficulty matching the tonal gradations and atmospheric effects of watercolor through the use of line only. Furthermore, the engraving medium required sharp definition where the watercolors could vaguely suggest forms, so the engraver often had to invent the necessary details. Lithography, with its ability to employ smooth, whole tones and washes and to add shadows and even tints, was more suited to reproducing watercolors, though at a greater expense than with wood engraving. The prints, especially the engravings, also had to be reduced to a smaller format than the originals.

J. J. Young seems to have had considerable latitude in redrawing some of the compositions, evidently making some changes for aesthetic reasons and others to improve accuracy. Some alterations may have derived from comments by Ives, Newberry, or others, but Young or the printmakers seem to have made other changes from pure whim. Individual catalogue entries detail the changes between watercolors and particular prints, but some generalizations can be made about them.

The printmakers used their own discretion about duplicating the artist's staffage (the incidental figures, foreground objects, and vegetation for a landscape) and often introduced their own ideas instead. The copy artists were in part inspired by the text, but they also may have been driven by a desire to express themselves artistically and make whatever aesthetic "improvements" they could get away with—an understandable impulse, given the boringly uncreative nature of copywork. Sometimes such changes altered an image's meaning. When Young added two Indians to the foreground of the frontispiece print after Möllhausen's *Steamboat Explorer (Chimney Peak)* (see cat. 6a), the change heightened the sense of a dramatic encounter between Western European science, technology, and "civilization" and "primitive" Native Americans. Thus the print became an effective piece of nationalistic propaganda, although Möllhausen's writings and watercolor confirm that this was not the artist's original intent.

Some changes may have been to satisfy the requirements of the printed report. Perhaps at Ives' instructions, the engraver removed snags in the river to emphasize the river's navigability, although these impediments were clearly shown in Möllhausen's original watercolors (see, for example, *Purple Hill Pass, View Towards South* [cat. no. 4]; *Riverside Mountain* [cat. no. 13]; and especially *Monument Cañon* [cat. no. 17]). Since Ives' *Report* omitted almost all reference to steamboat entrepreneur George Alonzo Johnson and to the *General Jesup*'s earlier voyage up the Colorado River with

Lieutenant White, all visual references were excluded, too: where Möllhausen's watercolor *Jessups Rapid* (cat. no. 28) showed the rock cairn that White and Johnson had erected to mark their Head of Navigation, the engraver removed this detail and retitled the print simply *Deep Rapid*, probably upon Ives' orders.

For a lithograph in Newberry's geological report (cat. 36a), Young redrew Möllhausen's view of the camp at Diamond Creek (cat. no. 36) from a higher vantage point to illustrate the lines of strata across the canyon's walls and to give a better idea of the granite rock in the foreground. However, the watercolor's lower vanishing point is more correct according to the rules of perspective, and the general topographical outlines shown in the original watercolor correspond more closely than the print with a recent photograph taken at the site. Thus, printmaking "corrections" did not necessarily improve on the original or entail fidelity to the actual topography.

Some changes were intended to correct inaccuracies in the watercolors, such as the impossibly twisted and distorted geological formations in *Painted Cañon*. However, in inserting more plausible landforms, the engraver also reduced the size of the steamboat and increased the height and magnitude of the canyon's walls and the mountains beyond, creating an exaggerated sense of scale in the print that renders the topography almost unrecognizable. Möllhausen's tendency to compress or telescope a number of picturesque topographical features into a single view also could make mountains and other formations appear taller than in reality. Usually these distortions increased as the engraver translated the composition into a print and sometimes further telescoped the topographical features, as in *Mouth of Mojave Cañon* (cat. no. 22). The engraver also cropped many of Möllhausen's compositions— another technique that tended to alter the original sense of scale (see *Monument Cañon*, cat. no. 17).

Young and the printmakers evidently read at least part of Ives' text, which led them to take considerable artistic liberties with Möllhausen's view of *Dead Mountain* (cat. no. 31). Ives' *Report* related the Mohaves' "superstitious" beliefs about a sacred mountain, located in Nevada's Newberry range. In the lithographed version (cat. 31a), the printmaker or copyist converted the simple topographical view into a romantic night landscape in which two silhouetted Indians in the right foreground gaze across the moonlit river toward the sacred mountain—transformed in the print into a smooth, featureless bald mountain with its peak enveloped in a mysterious haze.

The whereabouts of Möllhausen's original artworks from the time the printmakers copied them until they turned up in the brownstone in Brooklyn is not known. Perhaps Ives reclaimed them and passed them down through his family; he is known to have resided in New York both before and after the Civil War. Or the watercolors may have remained at the lithographer's in New York, eventually finding their way into the brownstone from there. Unfortunately, the mystery behind them may never be solved.

No finished watercolors have been located that correspond to forty of the prints in Ives' report or to any of the twenty-seven woodcuts in Newberry's geological report. Among these are seven lithographs of ethnographic subjects that credit Möllhausen as the original artist (see figs. 49-50).[210] These exemplify the standard Indian costume type genre, with two or three Indians arranged in a straight line across the picture plane, and may be composites of copy sketches that Möllhausen sent to Ives from Prussia, since individual figures in the lithographs resemble sketches in the artist's personal sketchbook. Only one full-page landscape lithograph credited to Möllhausen does not have a corresponding surviving original: *Mojave Cañon* (cat. 23a), which grossly distorts the topography and exaggerates its scale in relation to the steamboat.

Thirteen small-format woodcuts of landscapes in Ives' report could have been taken from missing Möllhausen originals.[211] Several contain elements characteristic of his style, including a sequence of birds taking flight in *Head of Mojave Cañon* (cat. 23c) and the genie-like bearded man sitting alongside a pipe-smoking Indian in *Ireteba's Mountain* (cat. 34b).[212] The first seventeen woodcuts in Newberry's geologic report could also be after Möllhausen's sketches, since the two men worked together until May 1, 1858, when the expedition split up to go in different directions at the Little Colorado River. Again, elements of several prints are similar to works by Möllhausen; for example, the Mohave raft and topography in *View of Northern Entrance to Monument Cañon* (cat. 19b) and the two small figures, one posed with a rifle and the other hammering away with what might be a geologist's pick, in *Hills of Tertiary Conglomerate and Modern Gravel, Camp 49* (cat. 28c).

Eight large, fold-out transfer-lithographed panoramas and seven full-page transfer lithographs in Ives' *Report* specifically credit Egloffstein, with whom Möllhausen shares the honor of being the first artist who definitely depicted the Grand Canyon.[213] These panoramas (see cats. 31d and 39a), with their long, horizontal format, often convey a better sense of the vastness of the Grand Canyon and the general topography along the Colorado than do Möllhausen's sketches. By selecting lofty prospects and emphasizing the general rather than the particular, Egloffstein partially solved the problem seen in many of Möllhausen's works—the fragmentation of general nature. Since Egloffstein's originals have disappeared, one can only surmise from the prints that he had a greater talent than Möllhausen for suggesting vast distances from high promontories but was weaker at drawing human figures (although, as with the prints after Möllhausen, Young's full-page transfer lithographs after Egloffstein may not reveal much about the artist's original work).

Egloffstein's greatest contributions to the Ives *Report* were his topographical maps, which he personally engraved upon steel plates from his field notes and those of Lieutenant Ives. They were among the finest maps produced in America up to that time.[214] In the introduction to the *Report*, Ives noted:

> The maps have been drawn directly upon the plates, which will obviate the ordinary expense for engraving. The style is partly new. The system of light and shade has been frequently adopted; but the application of ruled tints—by which the light sides of the mountains are relieved, and the comparative altitudes of different levels exhibited—is original, I believe, with the artist. The beautiful and effective representation of the topography is the best encomium both upon the style and its projector. The privation and exposure to which Mr. Egloffstein freely subjected himself, in order to acquire topographical information, has resulted in an accurate delineation of every portion of the region traversed.[215]

As the Ives expedition report was being printed in the United States, radical changes were taking place in Möllhausen's world. His mentors were passing on: Lichtenstein in 1857, Humboldt on May 6, 1859, Duke Paul in 1860, and Friedrich Wilhelm IV in 1861. A third son, Richard, was born to the Möllhausens in March 1861 (their second son had died in infancy in September 1856). Möllhausen turned his attention to his personal projects and to his burgeoning career as a writer. The earlier publication of *Diary of a Journey* and the public recognition that the king had granted him in his absence made him something of a celebrity in his native land. In demand both as a speaker and a writer, he gave lectures and talks before various organizations, such as the Berlin Zoölogical Museum and the "Minerva" Lodge of the Freemasons at Potsdam; the latter counted not only Möllhausen but also Humboldt and Friedrich Wilhelm IV among its members.

Encouraged by the reception he received, Möllhausen worked that much harder on the manuscript and illustrations for *Reisen*. He also submitted excerpts from it to various periodicals, partly to create interest for the forthcoming book and perhaps to beat out piratical or unauthorized copies, which remained prevalent in the absence of stringent copyright laws. As early as 1858, a report by Möllhausen on "The Colorado River of the West" appeared in the geographic journal *Zeitschrift für Allgemeine Erdkunde*, and other versions or excerpts were published in the magazine *Das Ausland* in 1859.[216] A French translation of various episodes from his works appeared in *Revue Germanique* in 1859. In the first half of 1860, Hermann Mendelssohn in Leipzig published a cheaper edition of Möllhausen's first travel account, without illustrations, under the title *Wanderungen durch die Prairien und Wüsten des westlichen Nordamerika*. . . . Those who missed the thirteen chromolithographs included in the first edition could purchase them in a separate edition for six Prussian thalers. In 1860 and 1861, respectively, the largest German illustrated family entertainment periodical, *Die Gartenlaube*, and its competitor, *Der Hausfreund*, published several literary sketches by Möllhausen, each with a woodcut illustration based upon the author's design.[217]

Möllhausen's diary of his third journey, *Reisen*, appeared in two editions in 1861: a deluxe two-volume version published by Herman Costenoble and a less expensive single volume published by Otto Pürfurst. Both editions were produced in Leipzig and contained twelve illustrations based on watercolors by the author. The medium of reproduction appears to be what might be loosely termed "toned xylographs" or wood engravings in two or three colors, but they actually may have been reproduced by transfer lithography, in which the engravings were transferred to lithographic stones so that more copies could be printed. Inscriptions on all of the prints credit A. Edelmann of Leipzig; some were carved by W. Aarland's Engraving Establishment ("X.A. von W.A. sc.") and others by the F. A. Bruckhaus Engraving Establishment. In the foreword to *Reisen*, Möllhausen credited "the talented Herr

Leutemann, an especially prominent animal illustrator in Leipzig," for interpreting his sketches and for reproducing them "so faithfully on a reduced scale."[218]

Five of the twelve images from Möllhausen's *Reisen* are quite similar to lithographs or woodcuts in Ives' *Report* and to corresponding watercolors in the Amon Carter Museum collection.[219] They provide concrete proof that the artist made personal copies of watercolors pertaining to the expedition, for the prints were produced in Leipzig after he had sent the other set to Ives in America. The chief difference between the Saxon prints and the American images is that the former—undoubtedly thanks to Leutemann—contain more animals, including bighorn sheep, pronghorn antelope, coyotes, eagles, cranes, and egrets or herons. Although both Saxon and American printmakers usually felt obliged to change Möllhausen's foregrounds, the *Reisen* prints are more faithful to the watercolors' more distant topography than are the American woodcuts. The same scientific aesthetic that influenced Möllhausen undoubtedly affected Leutemann and the German printmakers, who recognized the importance of topographical outlines. None of them wanted to be identified with the older European landscape tradition, castigated by landscape theorist Carus, in which ignorance of science and a propensity for conventionalized routine caused artists to alter the highly individualized contours of ridges so completely that "barely any trace remained of their peculiarly characteristic figures."[220]

Two of the prints in *Reisen* closely resemble Amon Carter Museum watercolors that were not reproduced in Ives' *Report*. These prints, modelled on copies of the watercolors, are generally faithful to the Amon Carter Museum's originals. However, the printmakers added various staffage in Möllhausen's foreground areas: *Schornsteinfelsen oder Chimney Peak* substitutes deer, birds, and beaver for the *Explorer*'s bow and skiff in the watercolor *Chimney Rock, from the East* (see cat. no. 7 and cat. 7b); the printed view of the mouth of Diamond Creek includes a campfire scene at the right which does not appear in the Amon Carter Museum's original (see cat. no. 38 and cat. 38b).

As Möllhausen prepared illustrations for *Reisen* and his various public presentations around 1859, he evidently began to paint watercolors that did not fit his usual genre categories of narrative scenes, landscapes, ethnographic costume pictures, or portraits of wild animals. From his field sketches he created "composite"

FIG. 70
After Balduin Möllhausen, *Eingeborene im Thale des Colorado (Natives in the Valley of the Colorado)*, lithograph, from *Reisen*, vol. 2, opp. p. 1

pictures that specifically illustrated Humboldt's scientific theories of geographical regions and ecological climate zones; four prints from *Reisen* and eight known watercolors formerly in the Berlin museums exemplify this composite genre.[221] One example depicts vegetation along the Colorado River Valley and desert (agaves, ocotillo, barrel cactus, saguaro, prickly pear, joshua tree, mesquite, cottonwood); another, vegetation of the plateau regions (junipers, firs, pines, oaks); and another, wild animals of the Colorado River region (deer, pronghorn antelope, and bighorn sheep).[222] Such scenes recall John Mix Stanley's edenic *Chain of Spires along the Gila* (1855), in which the artist crammed every manner of wildlife and vegetation onto a single canvas. Other Möllhausen compositions from c. 1859-60 depicted members of different Indian tribes living in the same geographic region. For example, the print titled *Eingeborene im Thale des Colorado (Natives in the Valley of the Colorado)* from *Reisen* (fig. 70) contains, from left to right, two Hualapai men; a Mohave woman, child, and warrior; two Yumas; a Chemehuevi woman and warrior; and an Apache warrior. In real life, such a

gathering could have easily resulted in violence, but Möllhausen arranged them together to illustrate the height and body build of average individuals in the various tribes.[223]

Beyond the geographical and scientific implications of such compositions, this genre had a practical benefit: it was easier and cheaper to produce one composite picture to illustrate multiple subjects than to produce separate works for each. For this reason, many of Möllhausen's original scenes showing individuals from only one tribe, as well as his individual tree studies or animal portraits, were never reproduced.[224]

LATER YEARS

Möllhausen's work for the Ives expedition concluded his career as an expedition artist, and he never again returned to the United States. However, the rich variety of his experiences in the 1850s fueled his burgeoning career as a novelist. In 1861, the same year that *Reisen* and his earliest *Hausfreund* articles appeared in print, Möllhausen also published his first novel, *Der Halbindianer* ("The Half-Blood"). This sentimental novel takes place primarily in the American West (Louisiana, Missouri, Kansas, California) and concerns "the prejudices of the Americans against every darker colored skin and the consequences arising therefrom." In this and many of his later novels, Möllhausen drew colorful sketches of people he had met or heard about in America and incorporated a considerable amount of American folklore. Through fiction, he was able to convey to the German public more impressions of American life than his factual travel journals and illustrations alone could do, and he continued to write novels and short stories about America until his death in 1905.[225]

Möllhausen's continuing position at the Potsdam royal libraries and his contacts in high social circles afforded him some security in this new career. Potsdam itself was a center for popular entertainment literature. In addition, as early as 1861 he became acquainted with Prince Friedrich Karl of Prussia, who in the 1870s began regularly inviting Möllhausen to Dreilinden, the prince's hunting lodge near Potsdam.

Although Möllhausen continued to paint and sketch, it became more of a personal therapeutic endeavor. In the 1870s, he described his yearly routine: "I write for eight to nine months, then

FIG. 71
Möllhausen in his study, 1905, from Preston Barba,
Balduin Möllhausen, The German Cooper (Philadelphia, 1914)

publication of *Reisen*. None of his oil paintings or the watercolors he made while on a trip to Norway with Prince Friedrich Karl in the summer of 1879 are known today.[227] Almost all of the surviving artworks concerning his three trips to America seem to have been created before 1861; although some of his views of buffalo, grizzly bears, pronghorn antelopes, and other animals could date from later, they more likely illustrated his lectures at the Berlin Zoölogical Museum in 1859.

Möllhausen's articles and fiction works occasionally contained illustrations, but these were frequently after sketches by other artists. When *Die Gartenlaube* ran several of his short stories in 1862, they accompanied illustrations by the Swiss painter Rudolf Friedrich Kurz (1818-1871), who had himself travelled extensively in the American West, even spending some time at Council Bluffs on the Missouri in 1851-52. When Kurz learned who would provide text for his pictures, he wrote to say he was "very glad" that Möllhausen had been selected rather than Otto Rupius, who "understands little or nothing about Missouri Indians and [who] would have ultimately embarrassed me."[228]

Möllhausen's transition to fiction-writing was propitiously timed, because developments in science were rapidly rendering his efforts in natural history obsolete. When Humboldt died in 1859, Charles Darwin was just publishing *On the Origin of the Species*, which would completely revise the organic but essentially static world view expressed in *Cosmos* and in Möllhausen's travel accounts. As a generalist in natural history, Möllhausen could not hope to keep up with the increasing specialization in the sciences, and the deaths of his mentors also meant the loss of associates to keep him informed of new developments in these fields. In addition, the American Civil War put a temporary halt to scientific expeditions there between 1861 and 1865. Such events made it a good time for Möllhausen to put his wandering days as a naturalist behind him and instead devote himself to writing novels.

Developments in the visual arts also must have influenced his decision. The rise of photography and of new methods for reproducing photographs was revolutionizing the way artists looked at the world and the way the public looked at art.[229] When the American government resumed its sponsorship of scientific expeditions after the Civil War, visual documentation increasingly became the province of photographers rather than topographical artists.[230]

for three to four months I permit myself to rest, that is, I occupy myself with watercolors and oils."[226] There is little evidence, however, of further artistic development on his part after the

Möllhausen was not the only one to see his work overshadowed by these changes: the post-Civil War explorations of the Colorado River by Major John Wesley Powell also eclipsed the achievements of the Ives expedition. The contributions of Ives and his men remain notable for a number of reasons, however. By determining the navigability of the Colorado River, the expedition helped to locate a supply route into Mormon Territory and to link this route with the Old Spanish Trail. Ives confirmed Whipple's idea that a good east-west railroad route could be found west of Cactus Pass in the Aquarius Mountains, and Ives' suggestion of Railroad Pass through the Cerbat Mountains near Kingman was the one used by the Southern Pacific.

The Ives expedition also had far-reaching consequences for the development of the Arizona Territory. The *Explorer* reached completely uncharted portions of the Colorado River, and soon steamers regularly plied the river to a point near Black Canyon. In the 1860s, the Colorado became a major highway for supplying the Territory's new capital at Prescott and the settlements that grew up when gold and silver were found in northern Arizona. LaPaz, Ehrenberg, and Hardyville became regular ports-of-call for the steamers, some of which reached as far as Callville (today under the waters of Lake Mead). Steamboating continued regularly on the river until the giant dams were built in the twentieth century.[231]

Ives' expedition also traversed territories belonging to several Native American tribes, visiting them without a major incident and without antagonizing them—despite Mormon attempts to turn the Indians against the U.S. government. As the first white Americans that several of the Indian tribes had seen, Ives and Möllhausen recorded crucial observations of native peoples just before white civilization's influence would change their lives forever.

And, of course, Ives and his men became the first whites to reach the depths of the Grand Canyon—an experience that had implications in several areas. The first scientifically trained people there, they could share the information they had gathered in a reasonably thorough, coherent, and accessible manner with the outside world. Dr. Newberry, as the first geologist to visit the Grand Canyon, formed what Goetzmann called "the most intelligent theory of the formation of the Colorado, or any western river, that had been made up to that time."[232] His tracing of a typical stratigraphic column of the Grand Canyon was "of fundamental importance in the development of American geology," and his

findings there helped geologists develop a more comprehensive view of the West.[233] Although John Wesley Powell later proved that some of Newberry's ideas were incorrect, Newberry's "first intelligent guess" was "for its time . . . contribution enough."[234] The geologist also explained that the Colorado Plateau, with its mesas and occasional igneous or volcanic intrusions, belonged, like the Grand Canyon of the Colorado, "to a vast system of erosion" that was "wholly due to the action of water."[235]

Baron Egloffstein's maps of the Colorado River and the Grand Canyon, based upon his own notes and sketches, as well as those of Ives, were incredibly fine pieces of work for their time and used a method of topographical shading in relief that was new for American mapmakers. Despite misjudgments about the location of the Little Colorado and the course of the Colorado River out of Utah, the maps were reasonably accurate in representing the topography that could be seen from the expedition route.

Egloffstein also shared with Möllhausen the distinction of being the first white man to sketch the Grand Canyon. Some of the prints from Egloffstein's panoramic sketches suggest that he had captured the character of the river's topography and that of the canyonlands better than Möllhausen. Egloffstein also was the first white man to visit Supai, and he alone produced some of the earliest images of the Hopi villages.

However, the baron's original sketches have never been located, leaving Möllhausen's sketches and watercolors from the Ives expedition as the earliest well-documented views of the Grand Canyon and other parts of present-day Arizona and California. Möllhausen also created some of the earliest visual records of Native American tribes such as the Chemehuevi and Mohave. While aesthetically pleasing, the sketches are not particularly remarkable from the standpoint of art history. The watercolors lack the drama and grandeur of the well-known canvases and sketches of Thomas Moran, and the painstaking geological accuracy of William H. Holmes' Grand Canyon views.[236] However, they are the first known views of the area, and considered with *Reisen* and Ives' *Report*, they form an excellent document of the Ives expedition and the earliest accurate accounting of the strange new lands they found. Möllhausen himself assessed the value of his watercolors in 1904, when he presented some of them to the Royal Kupferstichkabinett, which soon divided them between the Berlin Museum of Ethnology and the Berlin Geographic Society:

While they are not masterpieces, they nevertheless retain high value, since they come from times in which the almost completely extinct bison still wandered through the prairies by hundreds of thousands; and the disgracefully diminished natives still animated the western wilderness with splendid self-assurance. They form my most precious treasure.[237]

Möllhausen was quite accurate in this assessment, but he was far less prescient with one of his other observations, that the canyonlands they had visited "perhaps for centuries to come will remain a secret to mankind."[238] Ives similarly wrote: "Ours has been the first, and will doubtless be the last, party of whites to visit this profitless locality."[239] By publishing their works, however, Ives, Möllhausen, Egloffstein, and Newberry helped to begin a flood of scientific, artistic, and popular interest in the canyonlands that shows no sign of abating.

1 Correspondence between Harry L. Stern and Rick Stewart, ACM objects file 1988.1/1-47.

2 Samuel Chamberlain, a deserter from the army's dragoons, apparently made some views of a "Big Cañon" when he travelled to Arizona with a band of outlaws in 1849. Whether or not this was the actual Grand Canyon may never be proven, because Chamberlain often stretched the truth. The sketches were part of a manuscript titled *My Confession*, which dealt with Chamberlain's actual and fictitious exploits before, during, and immediately after the Mexican-American War of 1846-48. The manuscript and the sketches possibly date from the 1850s. See William H. Goetzmann, *Sam Chamberlain's Mexican War: The San Jacinto Museum of History Paintings* (Austin: Published for the San Jacinto Museum of History by the Texas State Historical Association, 1993) and his forthcoming reprint of Chamberlain's *My Confession*, from the illuminated manuscript at West Point.

3 *Reisen* is only now available in an English translation, by Dr. David H. Miller, forthcoming from the University of Arizona Press. Mollhausen's earlier travel account, *Tagebuch einer Reise vom Mississippi nach den Küsten der Südsee* (Leipzig: Hermann Mendelssohn, 1858), which dealt primarily with his second trip to the United States in 1853-54, appeared in English as *Diary of a Journey from the Mississippi to the Coasts of the Pacific with a United States Government Expedition*, trans. by Mrs. Percy Sinnett (2 vols.; London: Longman, Brown, Green, Longmans, & Roberts, 1858).

4 Works which assisted in this process included Horst Hartmann, "George Catlin und Balduin Möllhausen: Zwei Interpreten der Indianer und des Alten Westens," *Baessler-Archiv, Beiträge zur Völkerkunde* Neue Folge, Beiheft 3 (Berlin: Dietrich Reimer Verlag, 1963), and Robert Taft, *Artists and Illustrators of the Old West, 1850-1900* (New York: Bonanza Books, 1953), pp. 22-35, 278-284, especially pp. 25, 280, nn16-17.

5 There are two important biographies of Möllhausen. The best and most recent is Andreas Graf, *Der Tod der Wölfe: Das abenteuerliche und das bürgerliche Leben des Romanschriftstellers und Amerikareisenden Balduin Möllhausen (1825-1905)* (Berlin: Duncker & Humblot, 1994). Another biography, benefitting from contact with Möllhausen's widow, Caroline, is Preston Barba's *Balduin Möllhausen, The German Cooper*, Americana Germanica, vol. 17 (Philadelphia: University of Pennsylvania, 1914). Only toward the end of my research did I obtain a copy of Horst Dinckelacker's biography of Möllhausen titled *Amerika zwischen Traum und Desill-sionierung im Leben und Werk des Erfolgsschriftstellers Balduin Möllhausen (1825-1905)* (New York: Peter Lang Publishing, Inc., 1989), which is chiefly of interest for the interpretation of Möllhausen's literary works. David H. Miller's "Balduin Möllhausen, A Prussian's Image of the American West" (Ph.D. diss., University of New Mexico, Albuquerque, 1970) is the best source in English on Möllhausen's western travels.

Since Möllhausen's original manuscript diaries and notes are no longer extant, it is often virtually impossible to determine which came first: his written descriptions or his pictures.

6 The older Möllhausen wandered in America for several years, beginning in 1836, working variously as an architect and engineer in New Orleans (see Barba, *The German Cooper*, p. 37; Mary Louise Christovich, Roulhac Toledano,

et al., *New Orleans Architecture*, vol. 2 [New Orleans: Pelican Publishing Company, 1972], pp. 94, 228; New Orleans *Daily Picayune*, October 12, 1845, p. 4, col. 2). He produced at least one watercolor of a double house on Laurel Street, New Orleans, now in the New Orleans Notorial Archives; it reflects the same attention to detail that later characterized his son's drawing. He also attempted (apparently unsuccessfully) to secure a position as a major of artillery in the Army of the Republic of Texas. In 1848 Heinrich returned to Europe, where he was involved in an emigration scheme to colonize Virginia and Texas. He wrote a book on this subject, *Die in Texas und Virginien gelegenen, der Londoner allgemeinen Auswanderungs- und Colonisations-Gesellschaft gehörigen Ländereien*, which he published in Berlin in 1850, but by that time his son had gone to America. Heinrich eventually wound up as an engineer working on the construction of a railroad in the Crimea. He died in Odessa in 1867. See Graf, *Der Tod der Wölfe*, pp. 12-31.

7   Graf, *Der Tod der Wölfe*, pp. 36-37, 41. Graf cites report cards from the 1836-37 school year in the possession of the Möllhausen family, Bleicherode.

8   Möllhausen himself once wrote: "Orphaned early, at the age of 14 I was sent to Pomerania to become a farmer instead of being allowed to take things into account and become a painter." Short handwritten autobiographical note quoted in Dinckelacker, *Amerika zwischen Traum und Desillusionierung*, p. 2.

9   Ibid.

10 Johann Gottlieb Fichte, as quoted in translation by Gert Schiff, "An Epoch of Longing: An Introduction to German Painting of the Nineteenth Century," in *German Masters of the Nineteenth Century: Paintings and Drawings from the Federal Republic of Germany* (New York: Metropolitan Museum of Art, 1981), p. 10.

11 Miller, "A Prussian's Image," pp. 1-5; also David H. Miller, "A Prussian on the Plains: Balduin Möllhausen's Impressions," *Great Plains Journal* 12 (Spring 1973): 175-176, which cites the following quote from the journal of Swiss artist Rudolph Friedrich Kurz as particularly appropriate to Möllhausen's motivations: "From my earliest youth primeval forest and Indians had an indescribable charm for me. In spare hours I read only those books that included descriptions of the new world. . . I longed for unknown lands, where no demands of citizenship would involve me in the vortex of political agitations. I longed for the quietude. . . where neither climate, false modesty, nor fashion compels concealment in the noblest form in God's creation." See Rudolph Friedrich Kurz, *The Journal of Rudolph Friedrich Kurz: The Life and Work of This Swiss Artist* (Fairfield: Ye Galleon Press, 1969), pp. 1-2.

For an excellent general summary of European views of America, see Ray Allan Billington, *Land of Savagery, Land of Promise: The European Image of the American Frontier* (New York and London: W. W. Norton, 1981).

12 For an excellent discussion of the scientific aesthetic of the illustrated travel account, see Barbara Maria Stafford, *Voyage into Substance: Art, Science, Nature, and the Illustrated Travel Account, 1760-1840* (Cambridge, Massachusetts, and London, England: The MIT Press, 1984).

13 Most of these are bound in two sketchbooks belonging to Möllhausen's descendants in the town of Bleicherode, Germany. The Smithsonian Institution Archives and the Peabody Museum at Harvard University also retain some early works by the artist.

14 Duke Paul was a remarkable and unique individual, yet also typical of romantic travellers with an interest in art and the sciences. The nephew of the King of Württemberg, he held the rank of major general in the Prussian Army, a Ph.D. in medicine and anatomy, and was a member of several important European scientific societies. Disdaining the courtly life of European royalty, he preferred nature and the exotic. He had travelled extensively in the American Midwest and West, Mexico, the Near East, Algeria, Russia, and the Caribbean before he met Möllhausen. See Louis C. Butscher, "A Brief Biography of Prince Paul Wilhelm of Württemberg (1797-1860)" *New Mexico Historical Review* 17 (July 1942): 181-225; William H. Goetzmann, *Exploration & Empire: The Explorer and the Scientist in the Winning of the American West* (New York: Alfred Knopf, 1966), pp. 191-192.

15 Graf, *Der Tod der Wölfe*, pp. 56-70; Miller, "A Prussian's Image," pp. 13-59.

16 These would have been available either among the duke's baggage or in some of the various cities, towns, and settlements they visited together.

17 Duke Paul never completed and published a full account of the travels he made while with Möllhausen. The Stuttgart *Allgemeine Zeitung* (February 20, 21, 22, and 24, 1852, supplements 51-55) contains four continuing accounts of the Fort Laramie trip, which Duke Paul wrote and forwarded from Booneville, Missouri, on December 11, 1851. At the time the duke wrote this, Möllhausen was presumed lost on the prairie. Duke Paul's journals were valuable for their perceptive observations of America and for the few references they contained about Möllhausen. They remained in Württemberg after the Duke's death and eventually wound up in the Landesbibliothek in Stuttgart. Unfortunately, in World War II this library received a direct hit, and Duke Paul's journals were among the losses. However, a considerable number of notes and photographs of the journals were made before the war by Dr. Charles Camp, and the negatives are retained by the Lovejoy Library of Southern Illinois University at Edwardsville. Dr. David H. Miller has translated and edited them, and he kindly provided me with a photostatic copy of his work in progress.

Möllhausen's account of this expedition with Duke Paul is interspersed as "campfire stories" in his two published accounts of his later travels in America.

18 On Miller and Stewart, see Ron C. Tyler, ed., *Alfred Jacob Miller: Artist on the Oregon Trail* (Fort Worth: Amon Carter Museum, 1982). On Maximilian and Bodmer, see William H. Goetzmann, David C. Hunt, Marsha V. Gallagher, and William Orr, *Karl Bodmer's America* (Omaha: Joslyn Art Museum and the University of Nebraska Press, 1984).

19 Prince Alexander Maximilian von Wied-Neuwied, *Travels in the Interior of North America in the Years 1832-1834*, trans. by H. Evans Lloyd (London: Ackerman and Co., 1843; original German publication, Coblenz, 1839-41).

20 Duke Paul and Möllhausen certainly discussed Catlin's *Letters and Notes on the Manners and Customs of the North American Indians* (1841), which was the most oft-cited source on many North American tribes. A German translation of this work had appeared in 1848. The duke may have also introduced the young man to Catlin's *North American Indian Portfolio* (London: the author, 1844).

21 On Catlin, see William H. Truettner, *The Natural Man Observed: A Study of Catlin's Indian Gallery* (Washington, D.C.: Smithsonian Institution Press for Amon Carter Museum and National Collection of Fine Arts, 1979), and Brian

W. Dippie, *The Politics of Patronage: George Catlin and His Contemporaries* (Lincoln: University of Nebraska Press, 1991). For summaries of Catlin's achievements, see Goetzmann, *Exploration & Empire*, pp. 184-191, and William H. Goetzmann and William N. Goetzmann, *The West of the Imagination* (New York and London: W.W. Norton and Co., 1986), pp. 15-35.

22 Möllhausen's genre scenes recall the so-called "Biedermeier" style of painting which flourished in German-speaking lands between 1830 and 1850. The term is a compound of the names Biedermann and Bummelmeier, who were two characters from the satirical journal *Fliegende Blätter* (which in the 1860s and early 1870s published some of Möllhausen's stories). The word "Biedermeier" describes bourgeois smugness, or *Gemütlichkeit*, and it was first applied in art to cozy, unostentatious furniture based upon a pared-down version of the French Empire style. In painting it has been characterized as "a sunny, idyllic depiction of everyday life in a realistic style." One of its masters was Karl Spitzweg, whose name became a household word in Germany like America's Norman Rockwell. See Schiff, "An Epoch of Longing," p. 22.

23 Artist F. A. M. Retzsh popularized the style throughout German-speaking lands through his illustrations for the works of Goethe. See William Vaughan's chapter, "F. A. M. Retzsh and the Outline Style," in his *German Romanticism and English Art* (New Haven and London: Yale University Press for the Paul Mellon Centre for Studies in British Art, 1979), pp. 123-154, 279-281.

24 For a full account of these incidents, see Miller, "A Prussian's Image," pp. 49-54; Möllhausen, *Reisen*, vol. 1, pp. 137, 194-196; Duke Paul's diary for October 23 and 26, 1851.

25 Although Möllhausen sometimes blurred the distinctions between romantic fantasy and actual events in his early writings and sketches, and may have exaggerated details about his solitary adventure on the plains of Nebraska, there is little reason to doubt the general truth of it. Other sources confirm much of this incredible tale; for example, Swiss artist Rudolph Friedrich Kurz in his "Journal" entry for May 11, 1852, wrote: "Not long since, I am told, some Oto found, on the Platte, a Prussian named Mullhausen [*sic*] in a hopeless situation, having with him a wagon but no team. He is said to be an attendant of Duke Paul of Württemberg who was banished from court, and, so they say, he was protecting his Grace's silverware(?). Meantime where was the Duke?" The Council Bluffs newspaper *The Frontier Guardian*, January 9, 1852, p. 4, reported that "Paul William, Prince of Württemberg was picked [up] by Salt Lake stage about 235 miles from here. Four of his mules were frozen to death a few days before the stage came along." See Taft, *Artists and Illustrators of the Old West*, pp. 279-280. The Neues Palais Collection in Potsdam contains a finished watercolor of Möllhausen encamped alone in Nebraska.

26 Möllhausen, *Diary of a Journey*, vol. 1, p. 252. Other facts support this: Möllhausen had been horribly sick for most of the trip, some of his field sketches had been stolen, and mere survival must have demanded that he temporarily forget about sketching until he reached safety.

Bellevue's popularity for travellers along the Missouri River in the antebellum years is evidenced by the fact that Karl Bodmer and Maximilian had been there, as well as Swiss artist Rudolph Kurz. Remarkable for its candor is Möllhausen's surprising confession that while at Bellevue he forged paintings "in the Indian manner" on buffalo skins, for sale as souvenirs for Pierre Sarpy's tourist trade. See Möllhausen, *Diary of a Journey*, vol. 1, p. 291.

27 After Möllhausen sailed from New Orleans to Prussia in December 1852, Duke Paul continued his travels in the United States, and eventually made it to South America and Australia before his death in 1860. Unfortunately, the sketches Möllhausen had prepared for the duke remained unpublished at the latter's death, and many of them apparently were destroyed during World War II.

28 Merrill J. Mattes, "Robidoux Trading Post at 'Scott's Bluffs,' and the California Gold Rush," *Nebraska History, a Quarterly Magazine* 30 (June 1949): 125-126; and [Anon.], "Kansas History as Published in the Press," *Kansas Historical Quarterly* 17 (November 1949): 399.

29 Republic of Texas banknotes from the 1840s even had a buffalo hunt scene derived from Rindisbacher. See banknote, accession no. 1983.167, Amon Carter Museum, Fort Worth.

30 The best biographies of Humboldt in English are Douglas Botting, *Humboldt and the Cosmos* (New York: Harper & Row, Publishers, 1973), and Helmut De Terra, *Humboldt: The Life and Times of Alexander von Humboldt, 1769-1859* (1st ed.; New York: Alfred A. Knopf, 1955). The most comprehensive and scholarly is Hanno Beck's *Alexander von Humboldt* (2 vols.; Wiesbaden, 1959), in German.

31 For a survey of some of Humboldt's ideas connected with exploration, see William H. Goetzmann, *New Lands, New Men: America and the Second Great Age of Discovery* (New York: Viking Penguin Inc., 1986), pp. 52-60.

32 Although Humboldt was the most famous example, he was by no means the prototype of the universal scientist traveller in German-speaking lands. One of the most influential early examples was Georg Forster, an older friend of Humboldt's who had accompanied the English explorer Captain James Cook on his voyage around the world in 1772.

33 Lichtenstein to Raumer, April 2, 1853, quoted in Graf, *Der Tod der Wölfe*, p. 89.

34 Graf, *Der Tod der Wölfe*, pp. 88, 98-110, makes a strong case for this probability. The Möllhausen family still preserves some memorabilia from Humboldt's estate which have passed down through the generations.

35 Although Möllhausen's recommendations, personality, and his stories about America had the greatest effect in opening doors for him, his crude and naive sketches also drew praise and admiration. Part of this was due to the prevailing tastes in German art. For example, Gert Schiff in "An Epoch of Longing," p. 10, discusses the influential Protestant art theorist and German nationalist Wilhelm Heinrich Wackenroder, who longingly advocated the spirit of the Middle Ages—when he believed artists had been merely simple and untrained craftsmen, inspired only by piety and an honest mind—and wrote in 1797 that "[t]he German artist has no character, or he has the character of an Albrecht Dürer, Kepler, Hans Sachs, of a Luther or Jakob Böhme. Righteous, ingenious, thorough, precise, and profound is this character, thereby innocent and a little clumsy." Although probably unintentional on Möllhausen's part, innocence and clumsiness could certainly describe his artwork at this stage in his career.

36 It was Friedrich who refused to accept the Prussian crown offered him by the democratic Frankfurt Parliament in 1848, because he believed it was not divinely sanctioned. Friedrich's vacillation had precipitated the Revolution, which led to the disillusionment of thousands of liberal intellectuals and the

great wave of American emigration. Although Friedrich's conservatism proved a disaster in the long run, at the time he had managed to keep bloodshed to a minimum.

37 See Wilhelm Stegmann et al., *Deutsche Künstler in Lateinamerika: Maler und Naturforscher des 19. Jahrhunderts* (Berlin: Dietrich Reimer Verlag for the Ibero-Amerikanisches Institut, Preußischer Kulturbesitz, 1978). There are monographs in German on many of these artists, for example: Renate Löschner, *Otto Grashof: Die Reisen des Malers in Argentinien, Uruguay, Chile und Brasilien 1852-1857* (Berlin: Gebrüder Mann Verlag, 1987); Gilberto Ferrez, *Die Brasilienbilder Eduard Hildebrandts* (Berlin, DDR: Henschelverlag Kunst und Gesellschaft, 1989; also published in English as *The Brazil of Eduard Hildebrandt* (Rio de Janeiro: Distribudora Record de Serviçios de Imprensa, S.A., 1989); and Renate Löschner et al, *Bellermann y el paisaje venezolano 1842/1845: Bellermann und die venezolanische Landschaft, 1842-1845*, translated by Waltraud de la Rosa (Caracas: Asociacion Cultural Humboldt, 1977).

38 Letter of recommendation from Humboldt, dated Berlin, 16th of April, 1853, reproduced in Barba, *The German Cooper*, p. 160.

39 The best general summary of these expeditions is William H. Goetzmann's *Army Exploration in the American West, 1803-1863* (New Haven: Yale University Press, 1959), pp. 262-337. For a general summary of the artwork and information on the artists of the railroad surveys, see Taft, *Artists and Illustrators of the Old West*, pp. 1-35, 249-284.

40 While there were many precedents for exploration art, it may be significant that in the interval between the departures of Lewis and Clark's and Long's expeditions, Humboldt, the great proponent of on-the-spot scientific illustration, had visited the United States and met, among others, President Thomas Jefferson and Titian Peale's father, Charles Willson Peale, the artist, inventor, and founder of one of the first museums in America. Undoubtedly Humboldt's great atlas of 1811 and other illustrated volumes derived from his explorations in Latin America had provided the American scientific community with outstanding examples of expedition reports and illustrations. Moreover, his *Carte Generale du Royaume de la Nouvelle Espagne*—a map compiled from numerous Spanish sources in the viceregal archives in Mexico—remained for many years the most comprehensive gathering of data on the American Southwest. It was first published in Paris in 1811 in his two-volume *Political Essay on the Kingdom of New Spain*. See Goetzmann, *Army Exploration*, pp. 32-33.

41 There is an extensive bibliography dealing with the U. S. government exploration in the nineteenth century. Three of the most important are all by William H. Goetzmann: *Army Exploration* (1959); *Exploration & Empire* (1966, 1978); and *New Lands, New Men* (1986). These works also deal with some of the artists, as does Goetzmann and Goetzmann's *The West of the Imagination* (1986). For a more general introduction to the aesthetics of exploration art and the travel account, see Barbara Stafford, *Voyage into Substance*.

42 *Reports of Explorations and Surveys*, vol. 3, part 1, p. 3. Whipple had already worked as a topographical engineer on the Northwestern Boundary Survey from 1844-49, and as an assistant with the Mexican Boundary Survey from 1849-53. See Francis R. Stoddard, "Amiel Weeks Whipple," *Chronicles of Oklahoma* 28 (Autumn 1950): 226-230; Grant Foreman, ed., *A Pathfinder in the Southwest: The Itinerary of Lieutenant A. W. Whipple During his Explorations for a Railway Route from Fort Smith to Los Angeles, the Years 1853 & 1854* (Norman: University of Oklahoma Press, 1941), pp. 7-9.

43 Mary McDougall Gordon, ed., *Through Indian Country to California: John P. Sherburne's Diary of the Whipple Expedition, 1853-1854* (Stanford, California: Stanford University Press, 1988), p. 16, notes for comparison that the average salary of a male schoolteacher in Massachusetts at the time was "roughly $300 a year, and a bank cashier earned well below $1000." She states that "the New York *Daily Tribune* estimated in 1851 that an adequate income for an urban family was approximately $800."

44 A. W. Whipple to Möllhausen, May 10, 1853, quoted in Möllhausen, *Diary of a Journey*, vol. 1, p. ix. Despite Whipple's admonition, Möllhausen apparently kept no "official journal"; Miller, "A Prussian's Image," pp. 88-89.

45 Möllhausen wrote to Lichtenstein in Berlin on May 16, 1853 (see Graf, *Der Tod der Wölfe*, p. 92), stating that he had been hired as a draftsman, and that he had already started work making drawings from animal skins. He reported that the work was going better than he had anticipated. This was proof, Möllhausen wrote, "that a man can do what he truly wills to do." See also Möllhausen, *Diary of a Journey*, vol. 1, p. ix.

46 Miller, "A Prussian's Image," pp. 87-88; Muriel H. Wright and George H. Shirk, eds., "The Journal of Lieutenant A. W. Whipple," *Chronicles of Oklahoma* 28 (Autumn 1950): 241, 243.

47 For an in-depth analysis of the Whipple images, see my "Romanticism and the Scientific Aesthetic: Balduin Möllhausen's Artistic Development and the Images of the Whipple Expedition" (master's thesis, University of Texas at Austin, 1992). Also see Charles Evans, "Itemized List of the Whipple Collection," *Chronicles of Oklahoma* 28 (Autumn 1950): 231-234; Stoddard, "Amiel Weeks Whipple"; and Wright and Shirk, "The Journal of Lieutenant A. W. Whipple." It is amusing to note that not only did Möllhausen sign most of these sketches, but also some of Whipple's children or grandchildren—"Willie Whipple" and "Davie Whipple"—scrawled their names across some of them. The same collection also contains four sketches by Lieutenant John C. Tidball, a member of the military escort, several unattributed sketches, and several lithographs, of which some are proofs.

The surviving Whipple-related images may be divided into four general groups. Of the original sketches Möllhausen drew in America, sixty are in the Oklahoma Historical Society, having survived in a collection formerly belonging to Whipple's descendants. Many of these are field sketches, including views of forts, camps, settlements, towns, and landscapes in the Indian and New Mexican territories.

Nearly all of the sketches in the Oklahoma Historical Society's collection, whether finished or unfinished, served as models for a second group of images, the lithographs and engravings that illustrated official reports of the expedition. The third group consists of finished watercolors that Möllhausen, presumably working from his own set of copy-sketches (now lost), created back in Prussia between the fall of 1854 and 1857. Only a few of these watercolors survive in the Berlin Museum für Völkerkunde or are known through photographs taken before World War II; along with watercolors in the Neues Palais

Collection in Potsdam, they served as models for the fourth group of images: the chromolithographs and lithographs in the various editions of Möllhausen's *Diary of a Journey*.

There are also a few sketches by Möllhausen in the National Archives and the archives of the Missouri Botanical Garden in St. Louis. Many of the Whipple-related images by Möllhausen are known only through prints in the third volume of the *Reports of Explorations and Surveys to Ascertain the Most Practicable and Economic Route for a Railroad from the Mississippi River to the Pacific Ocean*, published by the U.S. government in 1856. Other related images appeared in 1857 in Möllhausen's published account of the expedition, *Diary of a Journey*. Many of the original paintings for this two-volume book remained in Berlin for many years and were tragically destroyed during World War II. Of the six originals in the Museum of Ethnology in Berlin that survived the war, only two or possibly three relate to the Whipple expedition. Fortunately, many of the lost paintings were photographed prior to the war. See the catalogue in Hartmann, "Catlin und Möllhausen," pp. 84-100; Taft, *Artists and Illustrators of the Old West*, pp. 25, 280.

48 As he had done earlier with Duke Paul, Möllhausen apparently made two sets of expedition images so he could give one set to his employers in Washington and take another set back to Prussia. This also safeguarded him from losing the entire collection in an accident. On several occasions during the course of the expedition, Möllhausen sent some of these sketches along with the collections and notes that he carefully packed and forwarded back east in the mail. On November 28, 1853, Prussian Ambassador Leo Gerolt wrote Humboldt from Baltimore that during his last visit to Washington he had seen "the beautiful drawings which Möllhausen had sent to the Smithsonian Institution, the originals of which he retains." Except for some of the botanical sketches drawn in graphite on tracing paper, it is virtually impossible and probably of little importance to tell which of the surviving sketches are actually the originals and which are the artist's copies. With the exception of the multi-media botanical sketches, generally the more finished works and several of the graphite sketches are attached within a ruled frame or border to paper mounts.

49 He applied watercolors in both a transparent wash style (mixing them with water only) and a gouache method (using both water and opaque white). For information on this medium, see Marjorie B. Cohn, *Wash and Gouache: A Study of the Development of the Materials of Watercolor* (Cambridge, Mass.: Fogg Art Museum, 1977).

50 Möllhausen signed most of the botanical sketches and dated them from July 17 to December 15, 1853. A few of the cacti are rendered on transparent tissue paper—perhaps because they are tracings from field sketches or because there was a scarcity of good-quality sketch paper toward the end of the expedition. (At Anton Chico, where the members of the expedition attended a fandango, Möllhausen wrote that he created "the most gorgeous shirt collars and fronts . . . out of stiff drawing paper"; *Diary of a Journey*, vol. 1, pp. 313-314.)

St. Louis botanist George Engelmann's contribution to *Reports of Explorations and Surveys*, 33rd Congress, 2d Session, Senate Exec. Doc. 78, vol. 4, part 5, p. 58, noted that a number of Möllhausen's cactus drawings, especially the sketches of "Cylindric Opuntiae" made on the spot during the expedition,

"greatly aided" the work of Paulus Roetter, who prepared the botanical illustrations for the report.

51 Catlin's prints pertaining to the landscape and tribes inhabiting southern and central Indian Territory were based on his sketches from an expedition of U.S. dragoons led by Colonel Henry Dodge, Lieutenant Colonel Stephen W. Kearny, and several other officers from Fort Gibson. See Truettner, *The Natural Man Observed*, and O. B. Jacobson and Jeanne d'Ucel, "Early Oklahoma Artists," *Chronicles of Oklahoma* 31 (Spring 1953): 122-126. Several volumes of Henry Rowe Schoolcraft's monumental *Historical and Statistical Information Respecting the History, Condition and Prospects of the Indian Tribes of the United States* (published in Washington from 1853-56 for the Bureau of Indian Affairs) were available to Möllhausen. They contained illustrations based in part on personal observations and eyewitness sketches by U.S. Army Captain Seth Eastman (1808-1875). See John Francis McDermott, *Seth Eastman, Pictorial Historian of the Indian* (Norman: University of Oklahoma Press, 1961).

Möllhausen's *Diary of a Journey* (vol. 2, pp. 11, 78, 134, 140, 274) praises former U. S. Boundary Survey Commissioner John Russell Bartlett's *Personal Narrative of Explorations and Incidents in Texas, New Mexico, Sonora, and Chihuahua*, 2 vols. (New York, 1854), which contained prints after Bartlett, Eastman, Henry C. Pratt, and others. Möllhausen did not see this book until after the Whipple expedition.

52 According to Gordon, ed., *Sherburne's Diary*, p. 17 n16, Whipple also carried Santa Fe trader Josiah Gregg's book *Commerce of the Prairie* (1844) and Randolph B. Marcy's *Report of the Route from Fort Smith, Arkansas, to Santa Fe, New Mexico*, Senate Exec. Doc. 64, 31st Congress, 1st Session (Washington, 1849-50). For summaries of the complicated movements and achievements of these and other explorers, see Goetzmann's *Exploration & Empire* and *Army Exploration*.

53 Lieutenant Colonel William H. Emory, *Notes of a Military Reconnoissance [sic] from Fort Leavenworth in Missouri, to San Diego, in California*, 30th Congress, 1st Session, Senate Executive Document No. 7 (Washington: Wendell & Van Benthuysen, printers, 1848). The Emory report prints included views of romantic ruins such as the "Catholic" and "Astek" churches at Pecos, the pueblo of San Felipe, other landscapes in New Mexico, Indian "hieroglyphics;" and zoological and botanical subjects such as cacti—all subjects similar or identical to Möllhausen's views of towns, Indian pueblos, ruins, and other landscapes. Stanley had already spent considerable time in Indian Territory since 1843, and on May 9, 1853, just a day before Whipple wrote to confirm Möllhausen's appointment to his railroad survey expedition, Stanley left Washington with Oregon Territorial Governor Isaac I. Stevens' survey of a northern route for the Pacific railroad. See Taft, *Artists and Illustrators of the Old West*, p. 14. The best source on Stanley is Julia Ann Schimmel, "John Mix Stanley and Imagery of the West in Nineteenth Century American Art" (Ph.D. diss., New York University, 1983).

54 James William Abert, *Journal of Lieutenant James W. Abert, from Bent's Fort to St. Louis in 1845*, Senate Ex. Doc. 438, 29th Congress, 1st Session (Washington, 1846); James William Abert, *Report of Lieut. J. W. Abert of His Examination of New Mexico, in the Years 1846-'47*, Senate Ex. Doc. 23, 30th Congress, 1st Session (Washington, 1848). For reproductions of Abert's original sketches see *Through*

the Country of the Comanche Indians in the Fall of the Year 1845; The Journal of a U.S. Army Expedition led by Lieutenant James W. Abert..., ed. John Galvin (San Francisco: John Howell Books, 1970), and *Western America in 1846-1847: The Original Travel Diary of Lieutenant J. W. Abert*, ed. John Galvin (San Francisco: John Howell Books, 1966).

55 As David J. Weber points out in *Richard H. Kern: Expeditionary Artist in the Far Southwest, 1848-1853* (Albuquerque: University of New Mexico Press for the Amon Carter Museum, 1985), pp. 219, 331-332 n22, Kern "was apparently still in Washington as late as May 22," since he put that date on a watercolor of Christ Church that belongs to the Amon Carter Museum. Kern was originally to have served as topographer and draftsman with Whipple because of his friendship with the lieutenant and his considerable experience along the proposed thirty-fifth parallel route. However, Kern had also served with explorer John C. Frémont in 1848-49 on an expedition to the Colorado Rockies, and Secretary of War Davis instead persuaded him to accompany the ill-fated expedition of Lieutenant John W. Gunnison along the proposed central railroad route between the thirty-eighth and thirty-ninth parallels. Gunnison and Kern left Washington toward the end of May 1853. See Möllhausen, *Diary of a Journey*, vol. 2, pp. 309-312; Weber, *Richard Kern*, pp. 237-240.

The printed profile and three-quarter view bust-length and standing portraits of Indians in James H. Simpson's *Journal of a Military Reconnaissance to the Navajo Country in 1849*, Senate Exec. Doc. 64, 31st Congress, 1st Session (Washington, 1850), after sketches by Kern and his brother Edward, resemble ones that Möllhausen drew on the Whipple expedition. On Edward Kern see Robert V. Hine, *In the Shadow of Frémont: Edward Kern and the Art of Exploration* (Norman: University of Oklahoma Press, 1982).

56 Lorenzo Sitgreaves, *Report of an Expedition down the Zuni and Colorado Rivers*, Senate Exec. Doc. 59, 32nd Congress, 2d Session (Washington, D.C.: Robert Armstrong, Public Printer, 1853 [facsimile reprint, Chicago: Rio Grande Press, 1962]). For an annotated account of Sitgreaves' expedition, combining several sources, see Andrew Wallace, "Across Arizona to the Big Colorado: The Sitgreaves Expedition of 1851," *Arizona and the West* 26 (Winter 1984): 325-364. As Gordon, ed., *Sherburne's Diary*, pp. 17-18, pointed out, no reference to Sitgreaves' report appears in Whipple's original manuscript journal of the railroad expedition, but Whipple and Möllhausen both mention Sitgreaves in their published accounts, which appeared in 1854 and 1857 respectively. See *Reports of Explorations and Surveys*, vol. 3, part 1, pp. 1-136, and Möllhausen's *Diary of a Journey*, vol. 1, p. 335. Möllhausen also mentioned artist Charles Preuss in his *Diary* and recounted John C. Frémont's exploits in some detail.

57 There is little information on John James Young, who also served as an expedition artist himself. The Washington, D.C., Census for 1870, dated June 3, 1870, p. 219, lists "Young, John J., dwelling no. 121, family no. 130, white male, age 40 at last birthday, occupation engraver, value of estate 5000, born in Prussia, father and mother of foreign birth, with a wife named Lavinia, age 29, who was a native of New York; a son John, age 12; a daughter Kate, age 4; and a baby named Oscar, age 2. All the children were born in Washington, D.C." Young probably anglicized his name when he emigrated to the United States. See Taft, *Artists and Illustrators of the Old West*, pp. 8, 34, 265-268 passim., 284;

Patricia Trenton and Peter H. Hassrick, *The Rocky Mountains: A Vision for Artists in the Nineteenth Century* (Norman: University of Oklahoma Press, 1983), pp. 102-103, 358.

That Möllhausen probably met Young during his second trip is evidenced by correspondence between Möllhausen and the director of the Smithsonian Institution, Dr. Spencer F. Baird. On March 13, 1855, Möllhausen, back in Prussia, wrote Baird to send his compliments to "Mr. Young on the Coast survey" (another project involving the Department of Explorations & Surveys). Still later, on May 12, Baird wrote him that "Mr Young has gone to North California with Lt. Williamson. They are to explore for a railroad pass between California and Oregon." Möllhausen to Spencer Baird, Potsdam, March 13, 1855, Smithsonian Institution Archives, Record Unit 53, Official Outgoing Correspondence, Assistant Secretary, 1850-1877, vol. 11 (reel 11), letter press books, p. 204.

58 Möllhausen did find time, however, to make a few sketches for Dr. Spencer Baird's child Lucy. Baird wrote Möllhausen on January 20, 1855: "My little Lucy holds you in lively remembrance treasuring the funny portraits you drew her." Smithsonian Institution Archives, Record Unit 53, Official Outgoing Correspondence, Assistant Secretary, 1850-1877, vol. 10 (reel 10), letter press books, p. 207.

59 Campbell and Tidball evidently gave Möllhausen some serious artistic competition, but how much influence they had on Möllhausen's work is more difficult to determine. Campbell had a good command of perspective and composition, and Tidball had a feel for landscape drawing, composition, perspective, and a fair ability to draw figures.

Campbell (1826-1899) at times commanded a separate detachment of the expedition, and he contributed several pages to Whipple's topographical report and six views which were lithographed or engraved. The print after his sketch of Albuquerque is the first known published view of that town. For biographical information on Campbell and a discussion of his works see Taft, *Artists and Illustrators of the Old West*, pp. 264-266. See also Gordon, ed., *Sherburne's Diary*, pp. 250; *Reports of Explorations and Surveys*, vol. 3, part 2, pp. 23-27; vol. 7, p. 23.

Lieutenant Tidball (1825-1908) commanded the military escort out of Fort Defiance, New Mexico Territory, which joined the expedition on December 12, 1853. He contributed three or four landscape sketches to the Whipple report. An 1844 West Point graduate, he had received some artistic training in topographic drawing from artist Robert W. Weir (1803-1889), who was drawing instructor at West Point from 1834 to 1876. For a short biography of Tidball, see Gordon, ed., *Sherburne's Diary*, appendix, pp. 263-264.

60 These included Thomas Sinclair (c. 1805-1881), who operated his lithography business in Philadelphia from 1838-81; A. Hoen & Co. of Baltimore, the longest-lived, continuously operating commercial lithographic company in the United States until it closed in 1981, and the lithographic partnership of Napoleon Sarony (1821-1896) and Richard C. Major, who operated their firm from 1846 until 1867. See George C. Groce and David H. Wallace, *The New-York Historical Society's Dictionary of Artists in America 1564-1860* (New Haven and London: Yale University Press, 1957), pp. 320, 558, 581; Nicholas B. Wainwright, *Philadelphia in the Romantic Age of Lithography* (Philadelphia: Historical Society of Pennsylva-

nia, 1958), pp. 49n, 89-90; Peter C. Marzio, *The Democratic Art: Chromolithography, 1840-1900: Pictures for a Nineteenth-Century America* (Boston: David R. Godine, 1979), pp. 49-51, 229; Lois B. McCauley, *A. Hoen on Stone: Lithographs of E. Weber & Co. and A. Hoen & Co., Baltimore, 1835-1969* (Baltimore: Maryland Historical Society, 1969); and Martha A. Sandweiss, Rick Stewart, and Ben W. Huseman, *Eyewitness to War: Prints and Daguerreotypes of the Mexican War, 1846-1848* (Fort Worth, Texas, and Washington, D.C.: Amon Carter Museum and Smithsonian Institution Press, 1989).

    Wood engraver Nathaniel Orr was active in New York as early as 1844 and in the 1850s was one of the country's leaders in his field. William Roberts (born c. 1829) was also active in New York as a wood engraver from about 1846 to 1876. "Pinckney" was perhaps Edward J. Pinckney or Samuel J. Pinckney (active 1848-1865), both New York City wood engravers. Benson J. Lossing (1813-1891) and William Barritt were both wood engravers and operated a firm in New York from 1847-69; Lossing was also an author whose most famous work, *Pictorial Field-Book of the Revolution*, was published in 1850-52. See Groce and Wallace, *Dictionary of Artists in America*, pp. 31, 404, 479, 507, 540.

61 Hartmann, *Catlin und Möllhausen*, p. 59, and Graf, *Der Tod der Wölfe*, p. 111, note that Humboldt had Möllhausen carry a letter to Catlin. Both sources state the letter is reproduced in Thomas Donaldson, "The George Catlin Indian Gallery in the U.S. National Museum" (Washington, 1886) for the *Annual Report of the Smithsonian Institution* (1885), p. 713.

62 Möllhausen to Spencer Baird, Potsdam, March 13, 1855, Smithsonian Institution Archives, Record Unit 7002, Spencer F. Baird Papers, 1833-1889, box 30.

63 Whipple to Humboldt, Washington, August 8, 1854, as quoted in Barba, *The German Cooper*, pp. 164-165.

64 Whipple et al., "Report upon the Indian Tribes," in *Explorations & Surveys*, vol. 3, part 3, p. 31.

65 Whipple et al., "Report upon the Indian Tribes," in *Explorations and Surveys*, vol. 3, part 3, p. 27.

66 Möllhausen to Spencer Baird, Washington, September 10, 1857, Smithsonian Institution Archives, Record Unit 52, Assistant Secretary, Incoming Correspondence, 1850-1877, Box 62, letter numbers 421 and 422. Note: two halves of this letter were incorrectly filed under two separate numbers.

67 I used an English edition of *Cosmos*, trans. by E. C. Otté (London: Henry G. Bohn, 1849). Vol. 2, pp. 440-457, contains the treatise on landscape painting, but Humboldt expressed his ideas on art in scattered references throughout his works. A useful summary by Renate Löschner, "Die künstlerische Darstellung Lateinamerikas im 19. Jahrhundert unter dem Einfluß Alexander von Humboldts," is in Stegmann et al., *Deutsche Künstler in Lateinamerika*.

68 Gert Schiff, "An Epoch of Longing," p. 15. For examples of Carus' works and ideas as well as other German romantics, also see Hubert Schrade, *German Romantic Painting*, trans. by Maria Pelikan (New York: Harry N. Abrams, Inc., 1977).

69 Carl Gustav Carus, *Neun Briefe über die Landschaftsmalerei*, quoted in Marcel Brion, *Art of the Romantic Era*, Praeger World of Art Series (New York and Washington, D.C.: Praeger, Publishers, 1966), pp. 110-111. The collected writings of Carus were published in Heidelberg in 1972 by Lambert Schneider,

in a facsimile edition of the expanded 1835 edition of *Briefe über die Landschaftsmalerei*, together with an Afterword by Dorothea Kuhn.

70 F. Arndt, *Eduard Hildebrandt, Der Maler des Kosmos* (Berlin: Verlag von R. Lesser, 1869), credited Hildebrandt with helping to establish a German school of watercolor painting. A native of Danzig, Hildebrandt had studied with French landscape painter Eugène Isabey in Paris before he came to Humboldt's attention in Berlin around 1843. Through the latter's recommendation, King Friedrich Wilhelm IV had sponsored Hildebrandt on trips to Brazil and the United States in 1844-45; to England, Scotland, the Canary Islands, Spain, and Portugal in 1847-49; to the central and eastern Mediterranean, including Italy, Greece, the Middle East, and Egypt in 1851-52; and to the Arctic in 1856. The exotic watercolor landscapes, portraits, and other studies that Hildebrandt produced from these journeys won him a membership and professorship in the Berlin Academy and the post of "court painter." For more recent information on Hildebrandt, see Ferrez, *Die Brasilienbilder Eduard Hildebrandts*.

71 Although the majority of these sketches in German collections were destroyed in World War II, enough of these works survive and enough appear in prewar photographs to make this assertion. As will be discussed later, the rediscovery of the watercolors for the Ives report helps confirm that a stylistic change in Mollhausen's works had occurred around 1855. Hartmann described the works from this period in German collections in *Catlin und Möllhausen*, pp. 59-100, but he merely grouped the pictures with Möllhausen's three trips and did not attempt to pin down the exact dates of execution of the finished sketches. The dates appearing next to the titles of some of these watercolor images evidently refer to when Möllhausen made his original field sketch or witnessed the scene or event, rather than the date he produced the final view. The stylistic similarities of these works to the watercolors in the Amon Carter Museum also support this idea.

72 In 1994 Mr. Kilian Klann, a German art collector, recognized the significance of the "lost" Möllhausen works in the Neues Palais Collection, Potsdam, and brought them to the attention of Möllhausen scholars. I am grateful to David H. Miller and Peter Bolz for sending me photocopies of the Potsdam watercolors in time to study them for this project. Unfortunately, the "discovery" of the Potsdam collection by western scholars was too late to include them in this exhibition. However, they have just become available in a catalogue: see Wilma Otte, Peter Bolz, et al., *Balduin Möllhausen—Ein Preussebei den Indianern. Aquarelle für Friedrich Wilhelm IV* (Berling-Brandenburg: Stiftung Preussische Schlösser und Gärten, 1995).

    Möllhausen recalled the three portfolios and the watercolors he had given to the King in his February 18, 1904, letter to the Director of the Berlin Kupferstichkabinett (quoted in Hartmann, *Catlin und Möllhausen*, p. 83). One of the watercolors was a portrait of the Mohave chief Mesikehota, whom Möllhausen met on the Whipple expedition. Writing from Washington to his wife, Caroline Möllhausen, on August 20, 1858, the artist mentioned that his third expedition had been saved from the danger of Mormon intrigues among the Mohaves through the efforts of "Chief Kairook, the same man with whom I travelled on foot several years ago, and Chief Mesikehota, whose portrait you find in the King's portfolios." Barba, *The German Cooper*, p. 173-174, reproduces this letter.

73 Humboldt to Alexander Mendelssohn (excerpt) Potsdam, 7. Nov. 1855, in the Staatsbibliothek Preußischer Kulturbesitz, West Berlin, Nachlaß Mendelssohn. I am very grateful to Dr. Andreas Graf for this important excerpt. The translation is mine.

74 Humboldt to Alexander Mendelssohn, Berlin, February 22, 1857, [SB Preußischer Kulturbesitz, Nachlaß Familie Mendelssohn, K. 1 A 1 d Nr. 119] in Graf, *Der Tod der Wölfe*, appendix, p. 291.

75 The Berlin firm of J. Storch & C. F. W. Kramer was founded in 1854 and actively produced chromolithographs and oleographs until 1879. They won medals for chromolithography at the World Expositions in Paris in 1855 and 1867, and in Philadelphia in 1876, and received numerous important commissions. Prior to this J. Storch ran his own business from 1836-1854. Winckelmann und Söhne was an important Berlin lithography firm and art dealership founded in 1828. They employed hundreds of hand colorers in the 1840s, exhibited chromolithographs at the Munich Industrial Exposition in 1854, and received an award at the 1855 World's Exposition in Paris. They were still in business as late as 1929, specializing in children's books. See Christa Pieske, *Bilder für jedermann: Wandbilddrucke 1840-1940* (München: Keysersche Verlagsbuchhandlung GmbH, 1988), pp. 30, 146, 150, 163, 181, 223.

    The lithography firm of Hanhart in London was one of the best in England. Michael Hanhart originally managed an English branch of the firm of French lithography pioneer Godefroy Engelmann, a student of Senefelder. Soon after this branch closed in 1830, Hanhart opened his own business. In the 1840s and 1850s his firm, sometimes styled M. & N. Hanhart, specialized first in printing tinted lithographs, then in chromolithography. See Michael Twyman, *Lithography, 1800-1850: The Techniques of Drawing on Stone in England and France and Their Applications in Works of Lithography* (London: Oxford University Press, 1970), p. 225.

76 A second German edition, appearing in Leipzig in 1860, was titled *Wanderungen durch die Prärien und Wüsten des westlichen Nordamerika vom Mississippi nach den Küsten der Südsee im Gefolge der von der Regierung der Vereinigten Staaten unter Lieutenant Whipple ausgesandten Expedition.* It contained only one lithograph and a map. The book also appeared in Dutch as *Reis van den Mississippi naar den kusten van den Grooten Oceaan*, trans. by H. D. Michaelis (Zuitphen, Netherlands: A.E.C. van Someren, 1858-59) and in Danish as *Vandringer gjuenem det verlige Nordamerikas prairier og udorkerner fra Mississippi til Sydhavets kyster*, trans. by M. Rorsing (Kopenhagen: Philipsen, 1862).

77 Taft, *Artists and Illustrators in the Old West*; Goetzmann, *Army Exploration in the American West*; David H. Miller, "The Ives Expedition Revisited: A Prussian's Impressions" and "Overland into Grand Canyon," *Arizona History* 13 (Spring 1972): 1-25 and (Autumn 1972): 177-196; Miller, "A Prussian's Image"; and Edward Wallace, *The Great Reconnaissance; Soldiers, Artists, and Scientists on the Frontier, 1848-1861* (Boston: Little, Brown, 1955) will remain indispensable references for a study of the Ives expedition and Möllhausen's role in it. Mollhausen's Ives expedition diary is forthcoming from University of Arizona Press.

78 Humboldt to U.S. Secretary of War Jefferson Davis, Berlin, March 24, 1857, translation in National Archives, Washington, D.C., Ayer Collection MS 395.

79 A native of New York City, Ives graduated from West Point in July 1852, served briefly in the Army's ordnance department, then transferred in 1853 to the Topographical Engineers. With Whipple's expedition along the 35th parallel, Ives led a small detachment from Fort Smith, Arkansas, to San Antonio, Texas, to obtain scientific instruments, supplies, and other needed equipment (including an inflatable rubber boat, which the expedition would use to cross the Colorado River into California). Ives' party did not join the main body of Whipple's men until they arrived in Albuquerque in October 1853. During the Civil War, Ives served as Jefferson Davis' chief aide when the latter was President of the Confederacy. He returned to New York City after the war and died there in 1868. See Frank Edward Ross, "Ives, Joseph Christmas," in Dumas Malone, ed., *Dictionary of American Biography*, vol. 5, part 1 (New York: Charles Scribner's Sons, 1932, 1961), pp. 520-521.

80 Ives to Möllhausen, n.d., as printed in Möllhausen, *Diary of a Journey*, vol. 2, p. 389.

81 Floyd to Ives, May 15, 1857, Records of the Office of the Chief of Engineers, RG77, National Archives, and Senate Exec. Doc. No. 11, 35th Congress, 1st session (Serial 920), *Report of the Secretary of War*, No. 2, "Report from Bureau of Exploration and Surveys, November 30, 1857," p. 39.

82 For a recent overview of Spanish exploration in the Southwest, see the first chapters of David J. Weber, *The Spanish Frontier in North America* (New Haven and London: Yale University Press, 1992). For information on Alarcón, see Herbert Eugene Bolton, *Coronado on the Turquoise Trail: Knight of Pueblos and Plains* (Albuquerque: University of New Mexico Press, 1949, 1964), and Alarcón's own account in George P. Hammon and Agapito Rey, eds. and trans., *Narratives of the Coronado Expedition, 1540-1542* (Albuquerque: University of New Mexico Press, 1940), pp. 124-155. On Garcés, see Elliott Coues, ed., *On the Trail of a Spanish Pioneer: The Diary and Itinerary of Francisco Garcés*, 2 vols. (New York: Francis P. Harper, 1900), vol. 1, pp. 161-162. In addition, Garcés made a daring journey from the Needles area to the Hopi villages, during which he visited present-day Kingman and Peach Springs and the rim of Havasu Canyon; Möllhausen would visit all of these areas except the Hopi villages during the Ives expedition.

83 The fort had been established by Major Samuel Peter Heintzelman in response to hostilities between Yumas and American civilians over the operation of a ferry; for the founding of Fort Yuma, see Arthur Woodward, *Feud on the Colorado* (Los Angeles: Westernlore Press, 1955), pp. 38ff, and Constance Wynn Altshuler, *Starting with Defiance: Nineteenth Century Arizona Military Posts* (Tucson: Arizona Historical Society, 1983). The government survey of the lower Colorado from the mouth to Fort Yuma was conducted by Lieutenant George Horatio Derby; see Odie B. Faulk, ed. and intro., *Derby's Report on Opening the Colorado, 1850-1851* (Albuquerque: University of New Mexico Press, 1969). Goetzmann, *Army Exploration*, provides an extensive discussion of American exploration along the lower Colorado.

84 See Sitgreaves, *Report of an Expedition down the Zuni and Colorado Rivers*; Andrew Wallace, "Across Arizona to the Big Colorado"; and Weber, *Richard H. Kern*, pp. 175-184.

85 Bolton, *Coronado: Knight of Pueblos and Plains*, pp. 138-142.

86 See *The Personal Narrative of James O. Pattie*, ed. by Timothy Flint (Cincinnati:

John H. Wood, 1831). For Army deserter Samuel Chamberlain, who claimed to have seen and sketched the canyon, see n. 2, above. Even if Pattie's and Chamberlain's claims are true, these previous visitors to the Colorado River's canyonlands did not leave notably accurate descriptions, and with the exception of Pattie's published narrative, their documentary efforts were inaccessible to most geographical researchers in the 1850s.

87 For information on the early steamboat pioneers on the Colorado River, see Richard E. Lingenfelter, *Steamboats on the Colorado River, 1852-1916* (Tucson: The University of Arizona Press, 1978). For a biography of George Alonzo Johnson, see Woodward, *Feud on the Colorado*.

88 Goetzmann, *Army Exploration*, p. 379; Miller, "A Prussian's Image," p. 148.

89 The Mormons also embarrassed federal troops by capturing several supply trains. The tardily dispatched federal troops could not reach Utah before winter set in, but instead spent it precariously at burned-out Fort Bridger in southwestern Wyoming. Fortunately, the "war" was resolved peacefully early in the summer of 1858. See Norman F. Furniss, *The Mormon Conflict, 1850-1859* (New Haven: Yale University Press, 1960).

90 Goetzmann, *Army Exploration*, p. 382. The Mormon situation was unfolding as Ives planned his expedition, but he may not have received instructions about this aspect of it from the War Department until late December 1857—after he arrived at the mouth of the Colorado River with the *Explorer*. On December 22, Captain Cadmus Wilcox brought Ives word of the Mormon War, along with two letters addressed from Captain Andrew A. Humphreys, Chief of the Office of Western Explorations and Surveys, warning of the possible Mormon move to Sonora and modifying Ives' original instructions. Also see Furniss, *The Mormon Conflict*, pp. 182-203.

91 Möllhausen wrote in *Reisen*, chapter 2, that he knew little concerning the expedition before he landed in New York except the date he was to depart from New York to San Francisco and "the general geographic direction we were to take."

92 Möllhausen to Spencer Baird, Berlin, June 12, 1857, Smithsonian Institution Archives, Record Unit 52, Assistant Secretary, Incoming Correspondence, 1850-1877, Box 62, letter number 420.

93 See *Report on the United States and Mexican Boundary Survey, Made Under the Direction of the Secretary of the Interior, by William H. Emory*, 34th Congress, 1st session, House of Representatives, Ex. Doc. No. 135, 3 vols (Washington: Conelius Wendell, Printer, 1857-1859; reprint, Austin: Texas State Historical Association, 1987) and *Survey of a Route for the Southern Pacific R.R. on the 32nd Parallel by A. B. Gray, for the Texas Western R.R. Company* (Cincinnati: Wrightson & Co.'s ["Railroad Record"] Print., 1856; reprint, Los Angeles: Westernlore Press, 1963). Interestingly, both of these publications were illustrated by German-speaking survey artists like Möllhausen: most of the Emory report's illustrations were after sketches by Arthur C. V. Schott (1813-1875), and the Gray report contained prints after sketches by Carl Schuchard (1827-1883). For more on Arthur Schott, see Gretchen Gause Fox, "Arthur Schott, German Immigrant Illustrator of the American West," M.A. thesis, George Washington University, 1977. For more on Carl Schuchard, see Robert Taft, *Artists and Illustrators of the Old West*, p. 269; Pauline Pinckney, *Painting in Texas: The Nineteenth Century* (Austin: The University of Texas Press for the Amon Carter Museum, 1967), p. 167; Robert Woodward, "Noted Artist Died at Corralitos: Charles Schuchard,

with A. B. Gray, Buried in Lonesome Grave," *Southwesterner* 4 (October 1964): 1, 3, 5, 9.

94 Humboldt to Möllhausen, cited in Barba, *The German Cooper*, pp. 159-160, and Graf, *Der Tod der Wölfe*, p. 124.

95 Möllhausen, *Reisen*, chapter 2. Möllhausen's connection with Corcoran is interesting. The latter travelled to Europe in the 1850s and probably met Möllhausen there. A personal friend of Alexander von Humboldt and of Baron Leo von Gerolt, the Prussian ambassador in Washington, Corcoran was also acquainted with Möllhausen's art instructor, Eduard Hildebrandt. Later, Möllhausen's wife wrote to Corcoran's daughter on at least one occasion. Upon Humboldt's death, Möllhausen offered to sell Humboldt's library to Corcoran, since a Prussian buyer was not forthcoming, and Henry Joseph of the Smithsonian Institution wrote to Corcoran of their interest in this purchase. See Joseph to Corcoran, Sept. 13, 1859, Library of Congress, Manuscript Division, reproduced in Graf, *Der Tod der Wölfe*, appendix, p. 293, and Henry Cohen, *Business and Politics in America from the Age of Jackson to the Civil War: The Career Biography of W. W. Corcoran* (Westport, Connecticut: Greenwood Publishing Corporation, 1971), pp. 100-101, 211, 290-291.

96 Möllhausen to Spencer Baird, Washington, September 10, 1857.

97 Under Lieutenants R. S. Williamson and H. L. Abbott, Newberry helped explore a railroad route from San Francisco Bay to the Columbia River. See George P. Merrill, "Newberry, John Strong," in Dumas Malone, ed., *Dictionary of American Biography*, vol. 7, part 1, pp. 445-446.

98 Baron (Freiherr) F. W. von Egloffstein was born in Altdorf, Bavaria, near Nuremberg, in 1824. The youngest son of Count (Graf) Wilhelm, Lord of the Manor Egloffstein in Franconia, F. W. served in the engineer corps of the Prussian army, and by 1850 he and his wife Irmgard had emigrated to the United States, settling in St. Louis. There he was a partner in a surveying and topographical firm. Apparently through the recommendation of the renowned St. Louis botanist, Dr. George Engelmann, Egloffstein served as topographer with John C. Frémont's fifth expedition (a privately funded railroad survey) in 1853-54. After becoming ill in the Rockies, Egloffstein left Frémont and joined Lieutenant E. G. Beckwith and the remnants of the Gunnison expedition (part of the U.S. Pacific Railroad Surveys), after Gunnison and several of his men were killed by Paiutes; Egloffstein was the replacement for expedition artist Richard Kern. Beckwith and his men explored a route northeast from Salt Lake City to Fort Bridger, then headed south of Great Salt Lake and west through northern Nevada and California. Evidently until the time of the Ives' expedition, Egloffstein was busy in Washington, D.C., working up his sketches and those of others for the maps and illustrations for the U.S. Pacific Railroad Surveys. In addition, he authored *New Style of Topographical Drawing* (Washington, 1857). On Egloffstein's later career, see n. 229, below.

The current Count Egloffstein (one of the the explorer's great-great-great-nephews) graciously invited Dr. Miller and me to visit the castle Egloffstein when we were in Germany in 1990. The count still has some of the letters that the explorer sent to his father, Count Egloffstein, concerning his early impressions of America and the Frémont Expedition. Also see Robert Taft, *Artists and Illustrators*, pp. 262-264; Goetzmann, *Army Exploration*, pp. 333-335, 393; Wallace Stegmaier and David H. Miller, eds., *James F. Milligan, His Journal of*

Fremont's Fifth Expedition, 1853-1854; His Adventurous Life on Land and Sea (Glendale, California: The Arthur H. Clark Company, 1988), pp. 266, 267; *Green's St. Louis Directory, for 1851* (St. Louis: Charles & Hammond, Book and Job Printers, 1850), p. 115; *The St. Louis Directory, for the Years 1854-5* (St. Louis: Chambers & Knapp, printers, 1854), p. 52; William I.. Montague, comp., *The Saint Louis Business Directory, for 1853-4* (St. Louis: E. A. Lewis, Printer, 1853, 1854), p. 49.

99 The Möllhausen family in Bleicherode, Germany, has many of the Ives-related works, including forty-two field sketches in Möllhausen's leather-bound book of around seventy unpaginated sheets, each approximately 14.1 x 20.3 cm in size. Until now, these works have never been reproduced in an English-language publication, and we are grateful to both the Möllhausen family and Dr. Friedrich Schegk of Munich for allowing us to reproduce them. Although American publications seldom make use of it, there is also Möllhausen material in the Berlin Museum of Ethnology, whose staff were most willing to assist us.

100 Möllhausen evidently began drawing in his personal sketchbook while at Fort Tejon, Mission San Fernando, Los Angeles, or possibly even at San Francisco. His sketch of bears could have been made in any number of places, including San Francisco, where he recorded seeing such animals along with their keeper and trainer, the famous "Grizzly Adams." German magazines published several unattributed California views between 1859 and 1861, along with descriptions by Mollhausen. These views included Missions San Fernando (*Der Hausfreund*) and San Luis Obispo (*Deutsches Magazin*).

101 Möllhausen, *Reisen*, chapter 3. Of course, Möllhausen could have observed camels back at the Berlin Zoo or in prints at the end of the trip; however, since the rest of the sketches in the sketchbook appear to have been drawn in the field, it is likely that he drew this one there as well.

102 Möllhausen to Spencer Baird, Aug. 29, 1858, Smithsonian Institution Archives, Record Unit 52, Assistant Secretary, 1850-1877, Incoming Correspondence, Box 62. Also see Möllhausen, *Reisen*, chapters 15 and 18.

103 Captain John N. Macomb's *Report of the Exploring Expedition from Santa Fe to the Junction of the Green and Grand Rivers* (Washington: Government Printing Office, 1876) contains a geological report by Newberry and excellent prints, including many fine chromolithographs by the Philadelphia lithographic firm of Thomas Sinclair & Son. These were apparently based upon J. J. Young's watercolor copies of Newberry's sketches. All of the images are topographical landscapes, but their style of execution is probably more Young's than Newberry's; for example, the unusual hats on the figures in the Macomb chromolithographs are similar to those in Young's watercolors in the National Archives.

104 National Archives, RG 77, Cupboard 3, shelf 5, books labeled "Colorado Survey." Although not signed, the original topographical maps in these three books were evidently the work of Lieutenant Ives and not Egloffstein, since they depict stretches of the river that Egloffstein did not personally see. Included in the original plats are sketches of the Colorado's delta and the area in Black Canyon beyond the Head of Navigation.

105 Ives, *Report*, part 1, p. 32.

106 Ives, *Report*, part 1, p. 34. Paul H. Taylor, one of the topographical assistants who returned downriver with the Explorer at the end of the river journey, wrote to Lieutenant Ives from Fort Yuma:

One of the trunks, containing photographic apparatus, I leave on board, it being so badly broken as to be unserviceable. I also leave with Captain Robinson, on board the steamer, where I think they will be better cared for than in the Q.M. storehouse, the photographic Camera; a box containing small quantities of acids; two rocket stands; a lot of tripods (including the broken camera tripod) a leather case of paper, and a small box, locked and strapped.

P. H. Taylor, Fort Yuma, April 24, 1858, to Ives, National Archives, Microfilm 506, roll 73.

107 Miller, "Balduin Möllhausen," pp. 160-161.

108 A finished watercolor made from one of these, titled *Das Gelände in der Nachbarschaft der Mündung des Gila in den Colorado* ("The terrain in the vicinity of the confluence of the Gila and the Colorado"), formerly belonged to the Berlin Geographical Society and was destroyed in World War II. See 1904 list, sketch no. 18. A lithographed view of Fort Yuma from the Carl S. Dentzel Collection, now belonging to the Phoenix Art Museum, has been attributed to Möllhausen. This, however, is probably the work of California artist and lithographer George Holbrook Baker.

109 The Cocopa were accustomed to visitors, having had contact with the Spanish as far back as 1540 and with Americans back to 1827. In the late nineteenth century the Cocopa actively participated in the steamboat traffic on the river. They supplied the steamboats with wood for fuel and were recognized as excellent river pilots and navigators. See Anita Alvarez de Williams, "Cocopa," in Sturtevant and Ortiz, eds., *Handbook of North American Indians*, vol. 10, pp. 99-112, and Lingenfelter, *Steamboats on the Colorado*.

Artist-naturalist Arthur C. V. Schott, while working with Major William H. Emory's Mexican Boundary Survey in 1853-55, had travelled in the vicinity and sketched the Cocopa a few years before Möllhausen. A colored lithograph after a Schott sketch of three Cocopas appeared in William H. Emory, *Report upon the United States and Mexican Boundary Survey*, vol. 1, opp. p. 111.

110 Ives, *Report*, part 1, pp. 31, 33.

111 Ives, *Report*, part 1, p. 39.

112 For information on the Quechan (Yumas), see Jack D. Forbes' well-written book *Warriors of the Colorado: The Yumas of the Quechan Nation and Their Neighbors* (Norman: University of Oklahoma Press, 1965), and Robert L. Bee, "Quechan," in Sturtevant and Ortiz, ed., *Handbook of North American Indians*, vol. 10, pp. 86-98. For information on Yuma-American relations, also see Arthur Woodward, *Feud on the Colorado*, pp. 19-30, and David P. Robrock, "Argonauts and Indians, Yuma Crossing, 1849," *Journal of Arizona History* 32 (Spring 1991): 21-40.

113 Möllhausen, *Diary of a Journey*, vol. 2, pp. 244-249. Whipple also noted this name. See also Möllhausen, *Reisen*, chapter 7; Whipple et. al, "Report upon the Indian Tribes," in *Explorations and Surveys*, vol. 3, part 3, p. 50n. An example of one of the war clubs, which Möllhausen collected while with Whipple, is in the Berlin Museum of Ethnology and is illustrated in Hartman, *Catlin und Möllhausen*, p. 81, fig. 1.

114 Woodward, *Feud on the Colorado*, pp. 87, 89, reproduces a photograph of Chief Pasqual, from which Julian Scott made his 1891 portrait for *Indians Taxed and Not Taxed....* Sturtevant and Ortiz, eds., *Handbook of North American Indians*, vol. 10, p. 94, reproduces another photograph of Pasqual, taken in the 1870s, as well

as other early photographs of Yumas. For information on Pasqual II, George A. Johnson, and the steamboat *General Jesup*, see Woodward, *Feud on the Colorado*.

115 Möllhausen, *Reisen*, chapters 7, 9, 10. For references to Maruatscha or "Capitan," as Ives called him, see Ives, *Report*, part 1, pp. 45, 70-71, 78; Möllhausen, *Reisen*, chapters 8, 11, 12, 14, 16.

116 Möllhausen, *Reisen*, chapter 9.

117 *Alta California*, May 7, 1859, and "The Yuma Indians," *Arizona Sentinel*, June 1, 1878, quoted in Woodward, *Feud on the Colorado*, pp. 86-88.

118 Johnson, "Exploration of the Colorado River," *Golden Era* (April 1888), p. 218, quoted in Miller, "A Prussian's Image," p. 162, and in Woodward, *Feud on the Colorado*, p. 76.

119 Ives, *Report*, part 1, p. 33, noted only Johnson's arrival at the delta and that he and his partner were "energetic proprietors."

120 Möllhausen, *Reisen*, chapter 8.

121 Ives, *Report*, part 1, p. 39; Andrew A. Humphreys to Ives, Nov. 18, 1857, copies, Letters Sent, Bureau of Explorations and Surveys, Record Group 48, quoted in Goetzmann, *Army Exploration*, p. 382.

122 Möllhausen, *Reisen*, chapter 8. There was some truth to these fears, as evidenced by the Mountain Meadows Massacre, in which Mormons purportedly incited Indians to kill some 120 members of an emigrant train in southwestern Utah in 1857. See Juanita Brooks, *The Mountain Meadows Massacre* (Stanford: Stanford University Press, 1950).

123 Ives, *Report*, part 1, pp. 45-46.

124 Egloffstein castle is beautifully located on a cliff overlooking the town of Egloffstein.

125 Möllhausen, *Reisen*, chapter 8; Ives, *Report*, part 1, p. 39. American stern-wheel steamboats were often compared to wheelbarrows in popular terminology.

126 Ives, *Report*, part 1, p. 53.

127 Ives, *Report*, part 1, p. 56.

128 Ives, *Report*, part 1, p. 62. Also see Goetzmann, *Army Exploration*, p. 386.

129 Möllhausen, *Reisen*, chapter 11. Möllhausen added that the only trouble with this was that it happened too frequently and soon lost its humor.

130 Goetzmann, *Army Exploration*, p. 384-385.

131 The Chemehuevi had seen other Europeans earlier, however; Spanish priests first documented contact with them in 1776. For information on the Chemehuevi, see Isabel T. Kelly and Catherine S. Fowler, "Southern Paiute," in William C. Sturtevant, gen. ed., *Handbook of North American Indians*, vol. 11, edited by Warren L. D'Azevedo (Washington, D.C.: Smithsonian Institution, 1986), pp. 368-397.

132 Möllhausen, *Reisen*, chapters 10, 12. Earlier, in *Diary of a Journey*, vol. 2, p. 244, Möllhausen, had written: "We had now before us, in great numbers, three different tribes of the natives, Chimehwhuebes, Cutchanas, and Pah-Utahs, who, however, did not differ at all in appearance."

133 Ives, *Report*, part 1, pp. 54-55.

134 Möllhausen, *Reisen*, chapter 10.

135 It is possible, though not necessarily likely, that the warrior on the right is the Chemehuevi man who took a ride on the *Explorer* on January 28. He had produced two old letters of introduction from Lieutenant Heintzelman, the former Yuma post commander, and from Lieutenant Whipple, the latter dated

February 22, 1854. Möllhausen then realized that he had sketched this man's portrait four years earlier. The man also remembered Möllhausen; Möllhausen, *Reisen*, chapter 10.

The Berlin Museum of Ethnology preserves a leather quiver full of arrows which Möllhausen brought back with him from the Colorado River area. It is of Chemehuevi or Mohave origin and is illustrated in Hartmann, *Catlin und Möllhausen*, p. 81, fig. 2. A photograph of a Chemehuevi cradle like the one in the lithograph is in Kelly and Fowler, "Southern Paiute," in *Handbook of North American Indians*, vol. 11, p. 379.

136 Goetzmann, *Army Exploration*, p. 386.

137 Maruatscha remained with the expedition until February 20. Möllhausen, *Reisen*, chapter 16.

138 Ives, *Report*, part 1, p. 68.

139 Möllhausen, *Reisen*, chapter 15.

140 Ibid.

141 Möllhausen, *Reisen*, chapter 17.

142 One recalls similar horseplay in William Sidney Mount's *Farmers Nooning*, in which a white boy tickles the ear of a young black man. See Alfred Frankenstein, *Painter of Rural America: William Sidney Mount 1807-1868* (Long Island, New York: The Suffolk Museum at Stony Brook, 1968), pp. 20-21.

143 Möllhausen, *Reisen*, chapter 17. Compare Ives, *Report*, part 1, p. 83. The reduced diet apparently had no effect on the Indians since they subsisted on other foods that the whites would not eat. See cat. no. 33.

144 Ives' cartographic sketch is in National Archives, RG77 (Office, Chief of Engineers) Field Survey Data, Cupboard 3, Shelf 5, "Colorado Survey, General Topography," No. 8, p. 54. Compare with Egloffstein's map 1.

145 Young had feared that Ives' steamboat might carry an invasion force. Incidently, the Mormon newspaper *Deseret News* attempted to keep its readers abreast of the progress of Ives' expedition. See Miller, "A Prussian's Image," pp. 186-189.

146 Möllhausen to Carolina Möllhausen, Washington, D.C., August 20, 1858, as printed in Barba, *The German Cooper*, Appendix II, pp. 173-174. Translation by David Miller.

147 See Arthur Woodward, "IratABA, 'Chief of the Mohave'," *Plateau* 25 (January 1953): 56-60; Ives, *Report*, part 1, p. 91.

148 It is important to note that Father Garcés, the intrepid Spanish missionary explorer, had probably ridden much of this same route on muleback on his way from the Needles area to the Hopi pueblos in 1776. He probably had only one or two Indian companions with him. See Eliot Coues, ed., *On the Trail of . . . Garcés*, vol. 2.

149 Möllhausen, *Reisen*, chapter 22.

150 Ives, *Report*, part 1, pp. 97-98.

151 Möllhausen, *Reisen*, chapter 23.

152 Goetzmann, *Army Exploration*, pp. 389-390, notes that "only Cardenas in 1541, Espejo in 1583, and Garces in 1776, and perhaps James Ohio Pattie and his fellow trappers in 1823 had even approached the canyon before, and they had not reached its depths."

153 Newberry in Ives, *Report*, part 3, p. 54.

154 Möllhausen, *Reisen*, chapter 22.

155 Ives, *Report*, part 1, p. 100.

156 Möllhausen, *Reisen*, chapter 7.

157 Möllhausen, *Reisen*, chapter 23.

158 Ives, *Report*, part 1, p. 101.

159 Möllhausen, *Reisen*, chapter 23.

160 Ives, *Report*, part 1, p. 104.

161 Möllhausen, *Reisen*, chapter 24.

162 Compare Ives, *Report*, part 1, pp. 104, 110; Möllhausen, *Reisen*, chapters 24 and 25; Egloffstein's map no. 2; and USGS 1:250,000 map "Williams." John G. Bourke mentioned "Forest Lagoons" in his diary of a visit there on November 14, 1884. See Casanova, ed., "General Crook Visits the Supais," *Arizona and the West* 10 (Autumn 1968), p. 276 and n. 69.

163 On their way back to the lagoons, the Ives party decided to follow the mysterious wagon trail that crossed their path. They found that it led northwest and then turned around, having been barred from further progress by the vast canyonlands. Subsequent investigations confirmed their suspicions that Lieutenant Beale had blazed the false trail. Ives, *Report*, part 1, p. 110; Möllhausen, *Reisen*, chapters 24 and 25.

164 Möllhausen, *Reisen*, chapter 24; Ives, *Report*, part 1, p. 105.

165 Compare Möllhausen's careful descriptions in *Reisen*, chapter 25, with John C. Frémont's description of a bumblebee on the continental divide, cited in Goetzmann, *Army Exploration*, p. 82. This same careful attention to detail characterized the writings of Möllhausen's mentor Humboldt, when the latter had written about buzzing insects during his ascent of Mount Chimborazo decades earlier.

166 Ives, *Report*, part 1, p. 107.

167 Ives, *Report*, part 1, p. 108. The Havasupai call themselves havasúwaipí, or blue-green water people. "Coconino," a Havasupai word, derives from the Hopi word for the Havasupai and Yavapai Indians, meaning "little water." It has been variously spelled Cosninas, Cosninos, Cojnino, and Coconinos. Europeans have sometimes called them Yampais and have confused them with the Yavapai. Inhabiting the same territory since at least the twelfth century A.D., the Havasupai were not as socially and geographically isolated as it might seem, since for centuries they carried on considerable trade with their Hopi neighbors to the east. See Douglas W. Schwartz, "Havasupai," in Sturtevant and Ortiz, eds., *Handbook of North American Indians*, vol. 10, pp. 13-24. Also see John Gregory Bourke's more detailed 1884 description of the Havasupai village in Casanova, ed., "General Crook Visits the Supais," pp. 253-276. Möllhausen had earlier written unfavorably about the Cosninos, whom he believed to belong to the Apache tribes. See Möllhausen, *Diary of a Journey*, vol. 2, pp. 165ff.

168 Möllhausen, *Reisen*, chapter 25.

169 Ibid.

170 Ives, *Report*, part 1, p. 113. They probably hit Whipple's tracks about three or four miles north of the present town of Ash Fork, Arizona.

171 Möllhausen, *Reisen*, chapter 25.

172 Ives, *Report*, part 1, p. 112.

173 Möllhausen, *Reisen*, chapter 26.

174 Ibid. The Whipple expedition had passed farther to the south. See Whipple, "Report," in *Explorations and Surveys*, vol. 3, part 1, pp. 88-89; part 2, pp. 32-34;

Möllhausen, *Diary of a Journey*, vol. 2, pp. 172-173, 176. See also Foreman, *A Pathfinder in the Southwest*, pp. 185, 186, and Will C. Barnes, *Arizona Place Names* (1935; reprint, Tucson: University of Arizona Press, 1988) p. 320.

175 Möllhausen, *Reisen*, chapter 26; Ives, *Report*, part 1, p. 112.

176 Möllhausen, *Reisen*, chapter 26.

177 The framed picture appears in the photograph of Möllhausen taken in his studio. Also see number "14. Ein Grizzlybär (hängt eingerahmt in meinem Arbeitszimmer)" on 1904 list of watercolors given to the Museum of Ethnology in Berlin. It should be reiterated here that Möllhausen had also seen Grizzly Adams' bears in San Francisco in late October 1857 (*Reisen*, chapter 1), and he was also a regular visitor at the Berlin Thiergarten. These works were not the artist's first attempts at drawing bears. Another sketch of three bears, today in the Smithsonian's Natural History Archives, relates to his earlier trips to the U.S. Bears also appear in the print from *Reisen* titled *Wild des Colorado Gebiets and Vegetation des Hoch-Plateaus.*

178 Möllhausen, *Reisen*, chapter 26.

179 His other renderings of pronghorns are in the Museum für Völkerkunde (watercolor titled *Pronghorn Angelope*), in the Möllhausen family sketchbook 2 (no. 15, *Antilope*), and in the Neues Palais Collection in Potsdam (watercolor, *Antilocarpa Americana*).

180 Möllhausen, *Reisen*, chapter 31, expressed disappointment at the results of Ives' exploration of the Hopi villages: "unfortunately they had only collected oral accounts about the isolated tribe. They had failed to bring back vocabularies of the still-unknown language or sketches of their cities." Perhaps Möllhausen had forgotten about the two images after Egloffstein of a Moqui village and the Moqui interior scene, which appeared in Ives' *Report*. On Hopi contact with Europeans, see the relevant articles in volume 9 of Sturtevant and Ortiz, eds., *Handbook of North American Indians.*

181 Had Möllhausen accompanied Ives' group, he too might have had a prominent topographical feature named for him. Newberry Mesa in Coconino County and Egloffstein Butte in Navajo County still appear prominently on modern topographical maps. See USGS 1:500,000 map "State of Arizona." However, this would have gone against Möllhausen's stated preference for using original Indian names for topographical features, and a Mount Möllhausen in the middle of the Navajo lands might not have pleased him.

182 Ives, *Report*, part 1, p. 117.

183 Goetzmann, *Army Exploration*, p. 391.

184 The Neues Palais Collection at Potsdam contains a triple full-length portrait watercolor of *Pedro Pino, der Governadore von Zuni, and Zwei Seiner Krieger* ["Pedro Pino, the Governor of Zuni, and Two of His Warriors"]. The same collection also contains watercolors of Zuni pueblo.

185 Möllhausen, *Reisen*, chapter 29.

186 Lieutenant Tipton arrested the alleged murderer and subsequently brought him in chains to Fort Defiance. The accused was later transported to Albuquerque by Ives to stand trial in a civil court. See Möllhausen, *Reisen*, chapter 30.

187 The Hopis had followed Ives and Egloffstein from their pueblos to Fort Defiance to lodge a complaint with the fort's commander, Major William T. H. Brooks, against their Navaho neighbors. Möllhausen was sympathetic to their plight and believed that the Americans were ignoring legitimate complaints.

188 Ever since the War with Mexico there had been troubles between the Navaho and the representatives of the American government, but in 1858 the situation was particularly bad. A drought in 1857 and the inexperience of Major Brooks (who had assumed command of the fort in November 1857) were but two contributing factors to the grazing dispute. Before this, it had been common practice for Navahos to raid their neighbors' fields and ranchos and for their neighbors to raid them, taking captives. This happened repeatedly despite (and sometimes because of) U.S. Army attempts to keep the peace. Then, in late May 1858, Brooks ordered Navaho headman Manuelito to remove his cattle from some pasture near the fort, claiming it was government property. Manuelito refused, claiming the land was his, and Brooks had his troopers slaughter the entire herd. In reprisal, one of Manuelito's followers killed Brooks' black slave boy, and soon a full-scale war ensued. See Frank McNitt, *Navajo Wars: Military Campaigns, Slave Raids, and Reprisals* (Albuquerque: The University of New Mexico Press, 1972), especially pp. 298ff.

189 Along the way they stopped in the town of Cubero (Covero) and Pueblo de la Laguna, both of which Möllhausen had sketched earlier with Whipple (see fig. 29). The Neues Palais Collection, Potsdam, has finished watercolors by Möllhausen of both subjects.

190 Möllhausen, *Reisen*, chapter 37.

191 Möllhausen to Spencer Baird, New York, August 11, 1858, Smithsonian Institution Archives, Record Unit 52, Assistant Secretary, 1850-1877, Incoming Correspondence, Box 62.

192 In 1858 the king was declared mentally unfit to rule, and his brother took over the throne as Wilhelm I. Friedrich Wilhelm IV remained alive until 1861. See Kurt Borries, "Friedrich Wilhelm IV," in *Neue Deutsche Biographie*, vol. 5, pp. 563-566.

193 Prussian Ambassador Leo von Gerolt to Alexander von Humboldt, July 20, 1858, reproduced in Barba, *The German Cooper*, p. 172, no. 22.

194 Möllhausen to Frau Möllhausen, Washington, 20 August 1858; reproduced in Barba, *The German Cooper*, p. 173.

195 Möllhausen, *Reisen*, chapter 38.

196 Möllhausen to Spencer Baird, New York, August 29, 1858.

197 Ives to Humphreys, March 22, 1859; Floyd to Howell Cobb (Secretary of the Treasury), March 22, 1859; and Cobb to Floyd, March 22, 1859, National Archives, RG 77, Records of the Office of the Chief of Engineers.

198 The personal set of sketches and the ones for *Reisen* were part of the group he bequeathed to the Berlin Kupferstichkabinett, which later were split between the Berlin Museum of Ethnology and the Berlin Geographical Society.

199 Möllhausen, *Reisen*, chapter 11.

200 Stafford, *Voyage into Substance*, p. 425.

201 Like these birds, cranes have straight beaks, long legs and necks which remain extended in flight, and tend to migrate in a straight line or "V." Sandhill cranes are still found in this vicinity. See Kenneth V. Rosenberg et al., *Birds of the Lower Colorado River Valley* (Tucson: The University of Arizona Press, 1991), pp. 171, 351; Roger Tory Peterson, *A Field Guide to Western Birds* (Boston: Houghton Mifflin Company, 1961), pp. 23, 92-93; and John Farrand, Jr., *Western Birds, An Audubon Handbook* (New York, St. Louis, and San Francisco: McGraw-Hill Book Co., Chanticleer Press Edition, 1988), p. 137. Incidentally, Möllhausen later titled one of his novels *Die Reiher (The Cranes)*.

202 A fine Möllhausen watercolor depicting three buffalo is in the Berlin Museum of Ethnology, and one of the watercolors in the Neues Palais Collection in Potsdam also shows buffalo.

203 The Berlin Museum of Ethnology also still preserves a Mohave two-feathered headdress brought back to Prussia by Möllhausen, like the one worn by the warrior pictured at the right in one of the Berlin paintings.

204 Another version of this composition exists in the Neues Palais Collection in Potsdam.

205 Möllhausen, *Reisen*, chapter 18.

206 Ives to Captain [A. A. Humphreys?], Santa Fe, June 2, 1858. MS, National Archives, RG 48, Office of Explorations and Surveys, Miscellaneous Letters Received, 1855-61, Box 3.

207 Ives to Captain A. A. Humphreys, Washington, Dec. 15, 1858, National Archives, RG48, Office of Explorations and Surveys, Correspondence concerning Requisitions and Accounts, 1854-61, Box 1. This estimate of expenses did not include a position corresponding with Möllhausen's, who was at work in Prussia on the project. For more on the report-printing process, see Ann Shelby Blum, *Picturing Nature: American Nineteenth Century Zoological Illustration* (Princeton, N.J.: Princeton University Press, 1993), pp. 158-209.

208 As noted in n. 60, above, Sarony & Major had done the lithography for the Railroad Survey reports, including prints after Möllhausen's sketches for the Whipple expedition. By 1857 their partnership had expanded to include artist and lithographer Joseph Knapp (active late 1840s-after 1892?), and just before the work on Ives' *Report* they had also produced plates for the U.S.-Mexican Boundary Survey. After Sarony retired in 1867, the firm continued as Major & Knapp. See Groce and Wallace, *Dictionary of Artists in America*, p. 558 and Marzio, *The Democratic Art*, pp. 49-51.

209 See Superintendent of Public Printing to Captain A. A. Humphreys, May 11, 1861; A. A. Humphreys to Jno., Head Engineer, March 23, 1861; J. J. Young to Thompson, May 15, 1861; A. A. Humphreys to M.O.D. Defrees, May 16, 1861, in National Archives, RG48, Office of Explorations and Surveys, "Correspondence Concerning Colorado & San Juan Expedition, 1861," Box 1.

210 The other Indian chromolithographs are: *Chemehuevis, Mohaves, Hualpais, Moquis,* and *Navajos.*

211 These prints are *Explorer's Pass, Remains of Grand Mesa in Chemehuevis Valley, Head of Mojave Cañon, Mount Davis, Meadow Creek, Ireteba's Mountain, Cerbat Basin, Colorado Plateaus from near Peacock's Spring, Side Cañons of Diamond River, San Francisco Mountain from Colorado Plateau, Mount Floyd, Valley of Fort Defiance.*

212 Given Möllhausen's high regard for Iretéba and his preference for keeping Indian names on topographical features, it is quite possible that the artist sketched himself and the Mohave guide next to this landmark.

213 In the fold-out lithographic panoramas, Egloffstein made his initial sketches atop elevations that permitted grand, sweeping vistas. His series of river views depicts the navigable portion of the Colorado from Fort Yuma to the Head of Navigation, and Egloffstein may have personally engraved the panoramas before they were transferred, apparently mechanically, to lithographic stones.

Although there is no cross-hatching or shading, he nevertheless conveyed a sense of aerial perspective by utilizing wide strokes of the burin in the darker, foreground areas. The long panoramic views recall Caspar David Friedrich's sketches of the island of Rügen and also the panoramic photographs of Solomon Nunes Carvalho, who served with Egloffstein on Frémont's fifth expedition.

Like the panoramic fold-outs, Egloffstein's transfer lithographs have dark répoussoir foreground areas that frame the picture and create a sense of grandiose scale; see *Cane Brake Canyon* (cat. 6d), *Big Cañon at Mouth of Diamond River* (cat. 38d), and *Moquis Pueblos* (fig. 63). Several woodcuts with similar characteristics are also probably after sketches by Egloffstein; see *Blue Peaks* (p. 118), *View north from Oraybe Gardens* (p. 125), and *Navajo Valley* (p. 129) in Ives, *Report*. Egloffstein's human figures seem incidental or unnecessary to the composition, except to establish a sense of scale, and in the Hopi interior view where they are prominent, the girls' features and proportions are quite awkward, probably evincing the baron's inexperience at this genre.

214 Carl I. Wheat, *Mapping the TransMississippi West, 1540-1861*, 5 vols. (San Francisco: The Institute of Historical Cartography, 1960), vol. 4, pp. 95-101, called Egloffstein "without doubt a genius."

215 Ives, *Report*, p. 6. I have never seen a copy of Egloffstein's rare *New Style of Topographical Drawing* (Washington, 1857). Some of its contents might be in the appendix of Ives' *Report* titled "Remarks Upon the Construction of the Maps," which explained Egloffstein's process in great detail, noting how it was an adaptation of, but different from, a French method of reproducing topographical maps. The topographical features were shaded in both the French and the new method as if light was falling upon them from an oblique angle as upon a plaster model or as the sun strikes the terrain in reality. But whereas, according to Ives, the French method depended upon toned paper and hand-applied light tints to produce the illuminated portions, Egloffstein introduced "fine parallel lines, drawn upon the plate with a ruling machine" to delineate the proportionally predominate areas of plains, leaving the highlighted areas blank and adding more lines to create heavily shaded areas.

216 Both periodicals had earlier published excerpts from his first travel account. Another excerpt from the forthcoming *Reisen* appeared in September 1859 under the title "A Story from the Western United States." See Graf, *Der Tod der Wölfe*, for the best list of Möllhausen's writings as well as documents relating to his biography.

217 "Scenes from the Folklife of New Orleans," "Mirage in the Desert," "The Prairie Fire," and "The Canalboat" in *Die Gartenlaube*; "Die Mission San Fernando in Kalifornien" and "Ein mexikanisches Lagerbild" in *Der Hausfreund*.

218 Möllhausen, *Reisen*, foreword. Heinrich Leutemann (1824-1905), an illustrator, painter, and writer who had studied at the Academy in Leipzig, was active as an illustrator for several German magazines with which Möllhausen was already or would later be associated as a contributing writer, including the *Gartenlaube*, *Illustrierte Zeitung*, and *Über Land und Meer*. On Leutemann, see Ulrich Thieme and Felix Becker, *Allgemeines Lexikon der bildenden Künstler von der Antike bis zur Gegenwart* (37 vols.; Leipzig: E. A. Seemann, 1907-1950; reprint ed., Leipzig: F. Allmann, 1964), vol. 23, p. 147.

219 The five *Reisen* prints are *Felsformation in der Nähe der Mundung von Bill Williams fork*; *Die Nadelfelsen oder Needles (von Norden gesehen)*; *Ende der Schiffbarkeit des Rio Colorado. (Aussicht aus dem Black Cañon.)*; *Der Diamant-Bach (Diamond Creek)*; and *Schluchten im Hoch-Plateau und Aussicht auf das Colorado Cañon*. The five corresponding prints in Ives' *Report* are *Shore on Lower Colorado* (lithograph); *The Needles (Mojave Range)* (lithograph); *Mouth of Black Cañon* (wood engraving); *Near Head of Diamond Creek* (lithograph); and *Side Cañons of the Colorado* (wood engraving).

220 Carus, *Neun Briefe*, pp. 27, 50, 108-109, 118-119, 135, 139-141, quoted in Stafford, *Voyage into Substance*, p. 456.

221 The composites from *Reisen* are *Vegetation der Kiesebene und des Colorado-Thales*, *Eingeborene des nördlichen Neu-Mexiko.*, *Eingeborene im Thale des Colorado*, and *Vegetation des Hoch-Plateaus*. From the 1904 list of watercolors destined for the Museum of Ethnology: *Choctaws, Chickasas u. Cherokesen, angesiedelt in Arkansas, Schawnees, Delawaren u. Hueckas, Grenzgebiet der Prärie, Moqui- und Zuni-Indianer, altmexikanischen Herkommens, Hualpais, Mohaves, Chimehuevies im Thal des Col., die anderen im Gebirge, Wild des Colorado-Gebietes, Navahoes, Moquis und Apatsches*. Composites from the 1904 list destined for the Geographical Society are *Vegetation der Coloradothäler* and *Wild des Coloradogebietes*.

The fact that Möllhausen's illustrations for *Reisen* are the first specifically dated examples of his use of composites suggests that two watercolors based on Möllhausen sketches from the Whipple expedition—a lost watercolor composition depicting Choctaws, Chickasaws, and Cherokees, and another showing Shawnees, Delawares, and Wacos—probably date from after 1859.

222 The second was a watercolor titled *Wild des Colorado-Gebietes*, which formerly belonged to the Berlin Museum of Ethnology and was destroyed in World War II. See Hartmann, *Catlin und Möllhausen*, p. 95.

223 Möllhausen's 1904 list of watercolors lists the original for this print as *Hualpais, Mohaves, Chimehuevies und Apatsches in ihren Größenverhältnissen. Mohaves u. Chimehuevies im Thal des Col., die anderen im Gebirge*. ("Hualapais, Mohaves, Chemehuevis, and Apaches in their relative heights. Mohaves and Chemehuevis in the Colorado Valley, the others in the mountains.")

224 It is possible that Leutemann and the printmakers first introduced the idea of combining several compositions in order to reduce printing costs for Möllhausen. Although this might have been initially to the printmaker's disadvantage, it would have satisfied their client's urge to provide his readers with as much comprehensive information as possible.

225 On Möllhausen's subsequent career, see Graf, *Der Tod der Wölfe*, and Barba, *The German Cooper*. For a study of his literature, see Andreas Graf, *Balduin Möllhausen, Geschichte aus dem Wilden Westen* (Munich: Deutscher Taschenbuch Verlag, 1995).

226 Möllhausen to Franz Brümmer, August 8, 1872, quoted in Graf, *Der Tod der Wölfe*, p. 148.

227 Barba, *The German Cooper*, p. 62.

228 Kurz to Möllhausen, July 1, 1862, quoted in Graf, *Der Tod der Wölfe*, p. 145 (my translation).

229 Interestingly, for a short time, Möllhausen's former colleague Egloffstein would be in the forefront of these developments. After his service with Ives'

expedition and the outbreak of the Civil War, Egloffstein, largely at his own
expense, organized and commanded what he hoped would be an elite corps of
German-Americans: the 103d Regiment of New York Volunteers. After seeing
considerable action in General Ambrose Burnside's 1862 campaign near New
Bern, North Carolina, Egloffstein was seriously wounded and breveted out of
service as a brigadier general the next year. Returning to New York City, he
edited *Contributions to the Geology and the Physical Geography of Mexico* (New York,
1864) and helped found the Geographical Institute. He became actively
involved in developing a halftone process based upon a patent he secured in
1865. One authority even called him "the inventor of the halftone," and for a
while after 1869, he operated his own "Helio Engraving Institute" in New York
City. In 1878 Egloffstein and his family left the United States, taking up
residence in Hosterwitz near Dresden, Germany. Egloffstein apparently ran
into difficulties and his health became an increasing problem. From 1879 until
his death on February 15, 1885, he and his wife resided in Dresden, Germany.
Egloffstein's widow, Irmgard, died in either Dresden or Hosterwitz near Pillnitz,
Saxony, on Oct. 6, 1903.

    See Wilhelm Kaufmann, *Die Deutschen im Amerikanischen Bürgerkriege
<Sezessionskrieg 1861-1865>* (München u. Berlin: R. Oldenburg, 1911), p. 494;
Taft, *Artists and Illustrators*, p. 264. For Egloffstein's patent, see U. S. Letters
Patent #51103, Nov. 21, 1865. For his efforts to develop the patent, see S. H.
Horgan, "General Von Egloffstein, the Inventor of Half-Tone," *International
Annual of Anthony's Bulletin* 9 (1897): 201-204. Also see Frederick W. von
Egloffstein, Claim for Invalid Pension, July 19, 1869, National Archives; U. S.
Consul John Y. Mason to Commissioner of Pensions, dated March 19, 1885,
Dresden; Widow's Pension File, National Archives.

230 On the subsequent government surveys and the relationship of surveys and
    photography, see Joni Louise Kinsey's excellent *Thomas Moran and the Surveying
    of the American West* (Washington and London: Smithsonian Institution Press,
    1992), which also contains a useful selected bibliography.

231 Faulk, ed. and intro., *Derby's Report*, p. 16. See also Francis H. Leavitt, "Steam
    Navigation on the Colorado River," *California Historical Society Quarterly* 22
    (March and June 1943), pp. 1-25, 151-174.

232 Goetzmann, *Army Exploration*, p.388.

233 Goetzmann, *Army Exploration*, p. 391.

234 Goetzmann, *Army Exploration*, p. 389.

235 Newberry, in Ives, *Report*, part 3, p.45.

236 On Moran and Holmes and the popularized of the Grand Canyon in America,
    see Kinsey, *Thomas Moran and the Surveying of the American West* ,
    William H. Goetzmann, "Limner of Grandeur: William Henry Holmes,"
    American West 15 (May-June 1979: 20-21, 61-63, and William H. Goetzmann's
    pamphlet, "William H. Holmes' Panoramic Art" (Fort Worth, Amon Carter
    Museum, 1977). In Germany, Möllhausen prefigured and was partially eclipsed
    as an artist and writer on America's natural wonders by Rudolf Cronau, a
    travelling journalist and artist for the magazine *Die Gartenlaube*. See Ruth Keller
    and Hans Lohausen, *Rudolf Cronau, Journalist und Künstler* (1855-1839)
    (Solinger: Bergischer Geschichtsverein, 1989).

237 Möllhausen to the Director of the Berlin Museum "Kupferstichkabinetts,"
    Berlin, February 18, 1904, quoted and translated by David Miller, *Balduin
    Möllhausen*, p. 276. The full letter in German is reproduced in Graf, *Der Tod
    der Wölfe*, p. 400.

238 Möllhausen, *Reisen*, chapter 25.

239 Ives, *Report*, part 1, p. 110.

Bibliography of Frequently Cited Works

Altshuler, Constance Wynn. *Staring with Defiance: Nineteenth Century Arizona Military Posts.* Tucson: Arizona Historical Society, 1983.

Barnes, Will C. *Arizona Place Names*, first published in *University of Arizona Bulletin* 6 (January 1, 1935), General Bulletin no. 2, Tucson: University of Arizona. Reprint edition, with intro. by Bernard L. Fontana, Tucson: University of Arizona Press, 1988.

Bonsal, Stephen. *Edward Beale, a Pioneer in the Path of Empire, 1822-1903.* New York and London: G. P. Putnam's Sons, 1912.

Conrad, David E. "The Whipple Expedition in Arizona, 1853-1854," *Arizona and the West* 11 (Summer 1969): 147-178.

Coues, Elliott, ed. *On the Trail of a Spanish Pioneer: The Diary and Itinerary of Francisco Garcés.* 2 vols. New York: Francis P. Harper, 1900.

Forbes, Jack D. *Warriors of the Colorado: The Yumas of the Quechan Nation and Their Neighbors.* Norman: University of Oklahoma Press, 1965.

Foreman, Grant, ed. and annot. *A Pathfinder in the Southwest: The Itinerary of Lieutenant A. W. Whipple During His Explorations for a Railway Route From Fort Smith to Los Angeles in the Years 1853 & 1854.* Norman: University of Oklahoma Press, 1941.

Goetzmann, William H. *Army Exploration in the American West, 1803-1863.* New Haven: Yale University Press, 1959.

Granger, Byrd Howell. *Arizona's Names (X Marks the Place).* Tucson: Falconer Publishing Co., 1983.

Gudde, Erwin G. *California Place Names: A Geographical Dictionary.* Berkeley and Los Angeles: University of California Press, 1949.

Ives, Joseph C. *Report upon the Colorado River of the West, Explored in 1857 and 1858 by Lieutenant Joseph C. Ives, Corps of Topographical Engineers, under the Direction of the Office of Explorations and Surveys, A. A. Humphreys, Captain Topographical Engineers, in Charge. By Order of the Secretary of War.* 36th Congress, 1st Session, Exec. Doc. No. 90. Washington: Government Printing Office, 1861.

Lesley, Lewis Burt. *Uncle Sam's Camels: The Journal of May Humphreys Stacey Supplemented by the Report of Edward Fitzgerald Beale (1857-1858).* Cambridge, Massachusetts: Harvard University Press, 1929.

Miller, David H. "Balduin Möllhausen: A Prussian's Image of the American West." Ph.D. diss., University of New Mexico, 1970.

Miller, David H. "The Ives Expedition Revisited: A Prussian's Impressions," and "The Ives Expedition Revisited: Overland into Grand Canyon," *Journal of Arizona History* 13 (Spring 1972): 1-25, and (Autumn 1972): 177-196.

Möllhausen, Baldwin. *Diary of a Journey from the Mississippi to the Coasts of the Pacific with a United States Government Expedition.* Trans. by Mrs. Percy Sinnett. 2 vols. London: Longman, Brown, Green, Longmans, & Roberts, 1858.

Möllhausen, Balduin. *Reisen in die Felsengebirge Nord Amerikas, vom Hoch Plateau von Neu-Mexico, unternommen als Mitglied der im Auftrage der Regierung der Vereinigten Staaten ausgesandten Colorado Expedition.* 2 vols. Leipzig: Hermann Costenoble, 1861. Some printed by Otto Purfürst.

Möllhausen, Balduin. *Tagebuch einer Reise vom Mississippi nach den Küsten der Südsee.* Leipzig: Hermann Mendelssohn, 1858.

Sturtevant, William C., gen. ed. *Handbook of North American Indians.* Volume 10, *Southwest*, ed. by Alfonso Ortiz. Washington, D.C.: Smithsonian Institution, 1983.

Taft, Robert. *Artists and Illustrators of the Old West, 1850-1900.* New York: Scribner, 1953.

Wallace, Andrew. "Across Arizona to the Big Colorado: The Sitgreaves Expedition of 1851," *Arizona and the West* 26 (Winter 1984): 325-364.

Weber, David J. *Richard H. Kern, Expeditionary Artist in the Far Southwest, 1848-1853.* Albuquerque: University of New Mexico Press, 1985.

Woodward, Arthur. *Feud on the Colorado.* Los Angeles: Westernlore Press, 1955.

Möllhausen's images for the Ives expedition have been assembled here into a roughly geographical sequence, tracing the expedition's route from Fort Yuma on the lower Colorado River to the lands east of the Grand Canyon. Each finished watercolor from the Amon Carter Museum collection has a separate catalogue number, and where applicable, related lithographs, engravings, field sketches, and modern photographs are included in catalogue entries. Several of these watercolors incorporate dates in their titles, but these dates probably refer to when the explorers arrived at a particular area rather than the moment Möllhausen painted the finished watercolor. Since he kept both a diary of the trip and a sketchbook, it should have been relatively simple for him to date the watercolors even if he made them months later.

The photographs, by the way, were made during a retracing of the Ives expedition route in 1990. It should be remembered that ordinary photographic lenses tend to make objects seem more distant; thus, just as Möllhausen and the engravers sometimes brought geographical details closer to the foreground than they really are, so the photographs tend to make features seem farther away than they would appear to the naked eye.

As we retraced the expedition's course, we found that distant geographical features were relatively easy to identify, but locating the vantage point where Möllhausen made his watercolor views was always difficult and sometimes impossible. Not only were some of his foreground elements rather insignificant in the vast territory he depicted, but natural changes, such as shifting channels in the river, have altered some perspectives, and modern improvements, including the building of several dams, have widened the river and raised its level in many areas, so that some of the distinctive features Möllhausen depicted are now underwater.

Cat. No. 1

Yuma Shoals (View from Fort Yuma up the River)

CAT. NO. 2

Left Bank 20 Miles from Fort Yuma

CAT. NO. 3

Purple Hill Pass, View Towards North

CAT. NO. 4

Purple Hill Pass, View Towards South

CAT. NO. 5

25 Miles from Fort Yuma, Up the River (Purple Hill Pass)

CAT. NO. 6

Steamboat 'Explorer' (Chimney Peak)

CAT. NO. 7

Chimney Rock, From the East

CAT. NO. 8

Chimney Rock, From the North

CAT. No. 9

Red Rock Gate

CAT. NO. 10

Sleeper's Bend [Sleeper]

CAT. No. 11

Lighthouse Rock

CAT. No. 12

Mohave Indians

CAT. No. 13

Riverside Mountain

CAT. NO. 14

Distant View of the Monument Range

CAT. NO. 15

Monument Mountains

CAT. No. 16

Corner Rock

CAT. NO. 17

Monument Cañon (Cañon Through the Monument)

Cat. No. 18

Left Bank, Four Miles South of Bill Williams Fork (Shore on Lower Colorado)

CAT. NO. 19

Four Miles North of Bill Williams Fork, View Down the River (Monument Range from the North)

CAT. NO. 20

Mount Whipple and Boat Rock

CAT. NO. 21

Distant View of the Mohave Range or Needles

CAT. NO. 22

Entrance of Mojave Cañon (Mouth of Mohave Cañon)

CAT. No. 23

Mohave Cañon 1

CAT. NO. 24

The Needles, View from the North

Cat. No. 25

Boundary Hill (Boundary Cone)

CAT. NO. 26

Beale's Crossing (Beale's Pass)

CAT. NO. 27

Obelisk Mountain, View Down the River (Pyramid Cañon)

CAT. NO. 28

Jessups Rapid (Deep Rapid)

CAT. No. 29

Cottonwood Valley

CAT. NO. 30

Painted Cañon

115

CAT. NO. 31

Dead Mountain. View From Round Island (Dead Mountain, Mojave Valley)

Cat. No. 32

Twelve Miles South [sic., North] of Round Island (Gravel Bluffs South of Black Mountains)

CAT. NO. 33

Head of the Navigation (Mouth of the Black Cañon)

Cat. No. 34

Second Camp After Leaving the Colorado (Railroad Pass)

CAT. NO. 35

Entrance of Wallpay Cañon [Peach Springs Canyon]

CAT. NO. 36

Camp on Diamond Creek

CAT. NO. 37

Mouth of Diamond Creek on Colorado River, View from the North

CAT. NO. 38

Mouth of Diamond Creek, Colorado River, View from the South

CAT. NO. 39

Ireteba's Turn

CAT. NO. 40

Character of the High Table Lands (Camp—Colorado Plateau)

CAT. NO. 41

Cañon Where Ives Went Down/ View From the Hieght [sic.] (Precipice Leading to Cataract Cañon)

CAT. NO. 42

Doctor's Rest

CAT. NO. 43

Cañon near Upper Cataract Creek (Side Cañons of the Colorado)

CAT. NO. 44

Bill Williams' Mountain

CAT. NO. 45

Tree

CAT. NO. 46

Fremont

Balduin Möllhausen, *Mohave-Indianer*, watercolor, 1858,
Museum für Völkerkunde, given by the artist in 1904

Balduin Möllhausen, *[Mohaves Playing Ring Game]*, watercolor, 1854,
Museum für Völkerkunde, given by the artist in 1904

Balduin Möllhausen, *Walapai Indians on Diamond Creek*,
watercolor, 1858, Museum für Völkerkunde,
probably given by the artist sometime after 1904

132

Balduin Möllhausen, *Navajo Indianer*, watercolor, 1853?,
Museum für Völkerkunde, Berlin

Balduin Möllhausen, *Grizzly Bears (Der graue Bär)*, watercolor and
gouache on paper, 1859, Museum für Völkerkunde, Berlin

CAT. NO. 1

Balduin Möllhausen
Yuma Shoals (View from Fort Yuma up the River)
Watercolor and gouache on paper
7⅝ x 11⅛ in. (19.4 x 28.3 cm)
Signed l.r.: "Möllhausen"
l.l. on mount: "N. 1."
l.c. on mount: "Yuma Shoals."
verso, center on paper mount:
"View from Fort Yuma up the River"
ACM 1988.1.3

Cat No. 1

Fort Yuma, situated across from the mouth of the Gila River on an isolated volcanic hill about a hundred feet above the Colorado, was, according to Ives, "not a place to inspire one with regret of leaving. The bareness of the surrounding region, the intense heat of its summer climate, and its loneliness and isolation have caused it to be regarded as the Botany Bay of military stations."[1] Möllhausen, who remained in its vicinity for three weeks in late December 1857 and early January 1858, declared more tactfully that "the best thing about Fort Yuma was the view which one enjoyed from its high vantage point during favorable weather."[2]

The watercolor titled *Yuma Shoals (View from Fort Yuma up the River)* depicts a portion of this panorama. The clearly distinguishable forms of the Castle Dome Mountains and the elevated vantage point suggest that Möllhausen made his original sketch for the scene from the hill at Fort Yuma, looking northeast. His description of the view in *Reisen* closely corresponds to his watercolor and to the actual view today:

One could see alternately mountains, deserts and water all around in a splendid panorama. One could follow the course of the Colorado for a long way to the north, until some jealous hills hid it from view, and to the south to the point where the horizon sank over the level valley. . . . to the north

and west one sees the gigantic pillars and towers of the Dome Mountains. Their strange shapes seemed like mirages in the desert . . . The rock known as "Capitol Dome" or "Ar-with-a-que" is the most striking because of its shape. Its walls are so perpendicular and uniformly round that from a distance it looks like a tower of gigantic proportions. Many formations like this of varying size are found in the mountain chain. The mountains extend from east to west in a broad arc, encompassing a panoramic view to the north. In this area one sees flat land, which can fairly be described as an arid desert. The banks of the Colorado and the Gila are thick with green shimmering willows, but their waters provide fertility to only a small portion of the several square miles which the fort overlooks."[3]

Although Möllhausen's watercolor shows the broad Colorado in the foreground, today the river, much reduced in volume from

1A
John James Young after Balduin Möllhausen, *Yuma Shoals,*
engraving (electrotyped?), from Ives, *Report* (1861), part 1, p. 46,
Amon Carter Museum Library

former times, arrives from farther to the east, out of the picture to the viewer's right. Möllhausen also brought the Castle Dome Mountains in the watercolor closer to the viewer than they actually appear (see cat. 1b).[4]

The watercolor apparently served as the model for an engraving (cat. 1a) for Ives' *Report*, which further increases the height of the mountains and alters the two figures in the left foreground. Instead of the two robed natives in the watercolor, the engraver substituted an Indian in a loincloth and what appears to be a Mexican with sombrero and serape. The engraver also somewhat misinterpreted Möllhausen's *jacal* or hut in the center foreground by elevating the pitch of the roof.

---

1   Ives, *Report*, part 1, pp. 42, 43-44. For information on Fort Yuma, see Altshuler, *Starting with Defiance*, pp. 67-72. Among the earliest depictions of the fort is the lithograph after Carl Schuchard titled *Fort Yuma, at the Junction of the Gila and Colorado Rivers*, in A. B. Gray, *Survey of a Route for the Southern Pacific Railroad on the 32nd Parallel* (Cincinnati, Ohio: Wrightson and Co.'s ("Railroad Record") Print, 1856), opp. p. 67, which also shows the *General Jesup*.

2   Möllhausen, *Reisen*, chapter 7.

1B
The Castle Dome Mountains in March 1990

3   Möllhausen, *Reisen*, chapter 7. Ives, *Report*, part 1, p. 48, calls the Castle Dome Mountains the "Dome Rock range." Möllhausen considered several ranges to the northeast, north, and northwest of Yuma to be part of the Dome Rock range, including the present Chocolate and Castle Dome Mountains of Arizona and Chimney Rock or Peak (Picacho Peak) on the California side of the river). These are not to be confused with the present-named Dome Rock Mountains east of Ehrenberg, Arizona. Chimney Peak is almost due north of Fort Yuma, and would be out of the picture to the left according to the orientation of the watercolor. See USGS 1:250,000 "El Centro" and "Salton Sea," and 15′ series "Laguna" and "Picacho Peak."

The Castle Dome Mountains extend north to south for twenty-five miles along the east side of U.S. Highway 95 in Yuma County. Granger, *Arizona's Names*, p. 124, says Garcés called them the Cerros del Cajon ("hills of bare sides") and that in 1859 their Mohave name was Sierra Mo-Quin-To-ara. She also notes that Garcés called Castle Dome Peak, elevation 3793 feet, Cabeza del Gigante ("head of the giant"). Viewed from a distance, the formation, which is hundreds of feet square, seemed to the soldiers at Fort Yuma to resemble the dome on the nation's capitol, and the "name quite easily and naturally shifted to Castle Dome. Among local Mexican and Spanish settlers the peak has long been called La Pelota ('the ball')."

4   Compare with USGS 1:250,000 "El Centro," and Egloffstein's map 1. The original field maps, mostly drawn by Ives, are in a series of notebooks in the National Archives, RG77. Egloffstein used these maps for his two printed maps, reproduced as foldouts in Ives' published *Report*. For information and maps on the historic shiftings and meanderings of the Colorado River at Yuma, see Don Bufkin, "Geographic Change at Yuma Crossing, 1949-1966," in *Arizona and the West* 28 (Summer 1986): 155-161.

For an earlier description of the area, see John Russell Bartlett, *Personal Narrative of Explorations and Incidents in Texas, New Mexico, California, Sonora and Chihuahua* (New York: D. Appleton & Company, 1854; reprint, Chicago: Rio Grande Press, Inc., 1965), vol. 2, pp. 158-160. Bartlett claimed that Henry C. Pratt made a panoramic sketch of the area at the junction of the Gila. Bartlett himself apparently made the sketch titled *Junction of the Gila and Colorado Rivers Looking up the Gila*, which looks east, from almost the same spot that Möllhausen made his original sketch. See Robert V. Hine, *Bartlett's West: Drawing the Mexican Boundary* (New Haven, Connecticut: Yale University Press for the Amon Carter Museum, Fort Worth, 1968), plate 32.

CAT. NO. 2

Balduin Möllhausen
Left Bank 20 Miles from Fort Yuma
Watercolor and gouache on paper
8½ x 11 in. (21.6 x 27.9 cm)
signed l.r.: "Möllhausen"
l.l.: "7"
l.c.: "L. bank 20 m. from Fort Yuma"
ACM 1988.1.7

Cat. No. 2

About twelve miles above Fort Yuma, the *Explorer* first encountered a pass into the Purple Hills, which Ives reported "are but a few hundred feet in height, and . . . though picturesque, by no means grand; but . . . an agreeable change to the broad monotonous flats which we have been surveying for so many weeks."[1] Dr. Newberry described the Purple Hills more scientifically—"composed of granite and mica slates, associated with which are purple porphyries and trachytes, in sufficient quantity to impart to them their prevailing color"—and noted that the granite of the hills, "yielding somewhat readily to the action of the elements . . . has formed slopes receding from the river, giving the pass an outline strikingly in contrast with that of most of the cañons cut in the porphyritic rocks higher up."[2]

The party named the narrowing of the river as they entered the hills "Explorer's Pass," after their steamboat, but this name does not appear on modern maps. Twentieth-century alterations to the course of the Colorado River just north of Fort Yuma have made identification difficult. Either Möllhausen or Egloffstein made the original sketch for the print in Ives' *Report* (cat. 2a), but this sketch has not been located.

The expedition made camp (no. 13) just beyond Explorer's Pass on January 12, having made fifteen miles that day according to Möllhausen's estimate. The following morning Möllhausen had time free "to investigate a series of low-lying rocky hills which rose from the water above reeds and willows nearby" and to observe "several distant rocky mounds,"[3] although it is difficult to tell

whether he was referring to the unusual formation in this watercolor. Unfortunately, there are no descriptions of the vicinity that particularly illuminate this view, but it appears to show the Arizona side of the river, probably between Laguna and Imperial Dams. Although Ives noted that the camp above Explorer's Pass was "the first grass camp yet seen on the river," Dr. Newberry remarked on the area's "beds of gravel deeply cut by washes and covered with the giant cactus, (*Cereus giganteus*) and other characteristic forms of the vegetation of the desert."[4]

---

1 Ives, *Report*, part 1, pp. 46-47.
2 In Ives, *Report*, part 3, p. 20-21.
3 Möllhausen, *Reisen*, chapter 9. Modern changes in the river channel complicate exact identification. See USGS 1:250,000 map "El Centro"; 15' series "Laguna Quadrangle"; and compare with Egloffstein's map 1. The distant hill at left in Möllhausen's picture may be Ives' Boot Peak.
4 Ives, *Report*, part 1, p. 47, and Newberry in Ives, *Report*, part 3, pp. 20-21. The river apparently used to follow the course of the present All American Canal through the valley .

2A
John James Young after Balduin Möllhausen, *Explorer's Pass*, engraving, from Ives, *Report* (1861), part 1, p. 47, Amon Carter Museum Library

CAT. NO. 3

Purple Hill Pass, View Towards North
Watercolor and gouache on paper
8 x 11¼ in. (20.3 x 28.6 cm)
signed l.r.: "Möllhausen"
l.l. on paper mount: "9"
l.c. on paper mount: "Purple Hill pass./view towards north"
ACM 1988.1.9

CAT. NO. 4

Purple Hill Pass, View Towards South
Watercolor and gouache on paper
8 x 11½ in. (20.3 x 29.2 cm)
signed l.r.: "Möllhausen"
l.l. on paper mount: "10"
l.c. on paper mount: "Purple hill pass view towards south"
ACM 1988.1.10

CAT. NO. 5

Balduin Möllhausen
25 Miles from Fort Yuma, Up the River (Purple Hill Pass)
Watercolor and gouache on paper
8 x 11 in. (20.3 x 27.9 cm)
l.l. on paper mount: "11"
l.r. on paper mount: "25 miles from Fort Yuma/ up the river"
ACM 1988.1.11

Cat. No. 3

Möllhausen made three views of what he and the other expedition members called "Purple Hill Pass," which apparently begins at present-day Imperial Dam and extends up to Fisher's Landing on Martinez Lake.[1] Möllhausen's *Purple Hill Pass, View Towards North* (cat. no. 3), which did not appear as a print in Ives' *Report*, may be the first in the sequence, although this is by no means certain. Möllhausen illustrates the steamboat's steady progress upriver, breaking the stillness of the pass, through watery reflections, opaque white ripples in the water, and birds in flight, suggested by tiny brushstroke Vs. He recorded that

> Along the right bank . . . impressive rock masses . . . towered precipitously and majestically overhead. The vegetation became sparse and eventually consisted only of small willow thickets in the mouths of the ravines and on sandy islands. Masses of variegated trachyte porphyry exhibited the characteristic appearance of colossal cacti. Some were shaped like candelabra. Others resembled solitary sentinels towering over the barren rocks, while some adhered to the slopes of the hills like round protuberances.[2]

Cat. No. 4

Dr. Newberry observed: "Toward the upper end of this passage the scenery becomes bolder; the hills higher and more craggy, showing considerable variety and contrast of color. The materials which compose them are porphyries, trachytes, and tufas, pink, purple, white, blue, yellow, brown, &c. The colors are all vivid, and obscured by no vegetation form a landscape very different from any before presented to our eyes."[3]

Möllhausen sketched *Purple Hill Pass, View Toward South* (cat. no. 4) from heights just above the expedition's next camp. He described this colorful scene at dawn on January 14, when

> the whole west bank of the Colorado was bathed in sunshine. Bluish shadows enveloped the east side of the Colorado and

4A
John James Young after Balduin Möllhausen, *Purple Hills*, engraving, from Ives, *Report* (1861), part 1, p. 48, Amon Carter Museum Library

[I]n regions where nature's own masterpieces abound, one is filled with awe and one's own artistic sensibilities are heightened. This was the sort of panorama through which we now planned to force our way. The sparkling river's surface stretched out into the distance before us with foaming eddies and black tree trunks. Jumbled rocks, in artistic and scenic arrangements, formed overhangs like stage scenery. They jutted out into the river which reflected their splendor and beautifully illuminated their soaring cliffs and towers. A narrow stretch of woods separated the rocks from these deceptive reflections in the water. The wide smooth surface of the river shimmered pure pale blue as it reflected the cloudless sky. Because of the vast distance, the color of the rocks changed gradually from a lovely red to a pale violet to a hazy blue. Farther away in a hazy distance, new mountain ranges appeared before us as if to prepare us for the constantly changing scenery.[5]

He also reported that the "Colorado is rich in beaver, but they do not live there in beaver houses as on smaller streams. Instead

jagged mountain ranges formed an incredible outline on the river's mirror-like surface. The faded red-and-violet cliffs were still wet from the night's rain. They appeared to have been painted with fresh colors, and the glaring contrasts lent a beautiful aspect to the landscape.[4]

Cat. No. 5

The engraver faithfully reproduced this sketch for Ives' *Report* as best he could in a single color print (cat. 4a), making only minor alterations by changing the direction of the smoke and omitting a few snags in the river.

The same morning, Möllhausen also sketched the pass while looking upriver (cat. no. 5), although his monochromatic rendering does not convey the colors he described in his diary. Noting that numerous obstacles had impeded their progress, so they were still "only 25 miles above Fort Yuma," he considered the view to the north:

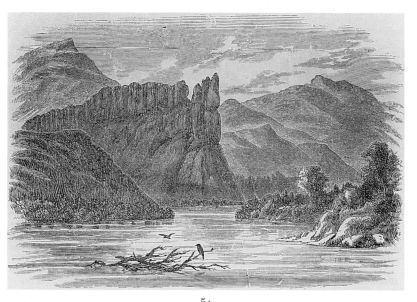

5A
John James Young after Balduin Möllhausen, *Purple Hill Pass*, engraving, from Ives, *Report* (1861), part 1, p. 49, Amon Carter Museum Library

they dig burrows in the banks of the river from which some tunnels open above and others below the water level. On long stretches of the steep clay banks, I repeatedly noticed such simple beaver burrows and many paths, as well as evidence of gnawed trees and branches."[6] He noted that Captain Robinson had set a beaver trap at their camp above Purple Hill Pass, and had snared one, although the wounded animal got away, dragging the trap with it.

---

1   Three ranges of purple hills cross the river in what are today part of the Chocolate and Laguna Mountains. Unfortunately, none of the inscriptions on the sketches or the print captions clearly distinguish between these ranges, and Laguna and Imperial Dams have altered the geography considerably. This complicates determining the exact identification and sequence of the views. Compare with USGS 1:250,000 map "Salton Sea" and 15' series map "Laguna Quadrangle" and Egloffstein's map 1.

2   Möllhausen, *Reisen*, chapter 9.

3   Newberry in Ives, *Report*, part 3, p. 21.

4   Möllhausen, *Reisen*, chapter 9. This camp, no. 14, was located on the California side of what is today the Imperial Reservoir, between Fisher's Landing and the mouth of Senator Wash. See USGS map "Salton Sea." Egloffstein's placement of Purple Hill Pass would be just above the mouth of Senator Wash.

5   Möllhausen, *Reisen*, chapter 9.

6   Ibid.

CAT. NO. 6

Balduin Möllhausen
Steamboat 'Explorer' (Chimney Peak)
Watercolor and gouache on paper
8⅛ x 11⅜ in. (20.6 x 28.9 cm)
signed l.r.: "Möllhausen"
l.l. on paper mount: "four 48 pieces"
l.c. on paper mount: "Steamboat Explorer."
verso, center on paper mount: "Steamboat Explora"
ACM 1988.1.1

Cat. No. 6

Lieutenant Ives noted that shortly after their party emerged from the Purple Hills on January 15, "the river swept around to the west, and soon entered a gorge or cañon more rugged and precipitous than any yet traversed." They called it Cane Brake Cañon for the cane that grew on the river's banks. Then, after a turn, they saw a prominent landmark that they had seen at a great distance from Fort Yuma—Chimney Peak, known today by the tautological name of Picacho Peak.[1] According to Ives:

> Its turretted pinnacles towered directly in front, and almost seemed to block up the head of the cañon. The vista was beautiful, and the channel looked promising. There was a fine head of steam on, and we anticipated making up at least ten miles before dark, when one of the rudder stocks broke. We were obliged to haul up to the bank to make a new one, and darkness came on before this was accomplished.[2]

Möllhausen probably combined at least two field sketches (now presumably lost or destroyed) to construct this finished watercolor. Camped around present-day Island Lake in the Imperial National Wildlife Refuge (camp no. 16) while the rudder was being repaired, he apparently made his original topographical sketch looking southwest toward the present settlement of Picacho in California's Picacho State Recreation Area.[3] He may also have sketched the *Explorer* at this time, although he could have done so at many points during the expedition.[4]

In both the watercolor and the print after it in Ives' *Report* (cat. 6a), the *Explorer*'s boiler with its prominent smokestack sits amidships, while the large pistons run past the cabin to the big wheel and narrow wheelhouse in the stern, so "the whole structure looked a little like a wheelbarrow." A six-foot-long wooden shack located toward the stern, "just wide and high enough barely to provide adequate space for more than three persons engaged in writing," has three figures on top: Captain Robinson leans on the curved tiller, while the man in front of him, possibly Dr. Newberry, cleans his rifle, and another figure sits and sketches—perhaps the artist himself.[5] Lieutenant Ives clarified the role of others on board:

> While the boat is in motion a man is stationed at the bow with a sounding pole, and constantly calls out the depth of the water and the character of the bottom. This is not so much for the benefit of the pilot as to gratify the anxious curiosity of the passengers, and to enable Mr. Bielauski [*sic*] and Mr. Egloffstein, who sit on the wheel-house with their notebooks

6B
Balduin Möllhausen,
17a. Sailor with a Pipe
[*Seeman mit Pfeife*],
graphite on paper,
from Sketchbook 3,
Möllhausen Family,
Bleicherode, Germany,
photograph courtesy of
Dr. Friedrich Schegk,
Munich

> delineating the river and the surrounding country, to keep an accurate record. . . . The working party remains near the bow, and the others distribute themselves as they best can over the limited accomodations afforded by the wood piles on either side of the boiler. What little space is left abaft the boiler, when the luggage is all aboard, is taken up by the fireman and by Mr. Carroll."[6]

According to Möllhausen, Lieutenant Ives and Lieutenant Tipton, the commander of the escort, sometimes joined hydrographer Bielawski and topographer Egloffstein on the bench above the wheelhouse. The Indian in a blanket next to the boiler may be one of the guides, Maruatscha or Mariano.

Carroll, the engineer, is also the subject of an existing field sketch (cat. 6b), which shows how he regulated the boat's speed by opening and closing valves on the steam pipes that supplied the pistons. In the sketch, Carroll wears a seaman's cap and smokes a long-stem pipe, while Möllhausen himself sits partly obscured by a wall of the shack. In the watercolor, Carroll (with his pipe) appears on the porch of the cabin, below Möllhausen and Newberry, and to the left of the fireman stoking the fire of the boiler.

6A
John James Young after Balduin Möllhausen, *Chimney Peak*,
toned lithograph (black with ruled gray), from Ives, *Report* (1861),
frontispiece, Amon Carter Museum Library

Comparison of Möllhausen's watercolor with the often reproduced lithograph (cat. 6a) illuminates how John James Young redrew, altered, and translated the scene for reproduction in Ives' *Report.* The figures in both the print and the watercolor may be out of scale with the steamboat, which Möllhausen noted had an iron-plate hull fifty feet long, ten feet wide, and four feet deep. Young's steamboat is longer and larger than Möllhausen's, and its smokestack more graceful. The mast near the wheelhouse patriotically bears the Stars and Stripes. Young's figures are smaller and more delicately proportioned; they wear wide-brimmed hats with almost conical crowns, and the man at the bow assumes a brave, heroic stance, staring straight upriver into the wild unknown instead of cautiously gauging the depth of the water with the sounding pole. This classically posed, confident figure recalls countless popular, almost clichéd images of Manifest Destiny and progress, including Emanuel Leutze's *Westward the Course of Empire* or *Washington Crossing the Delaware.* Möllhausen's figures, by contrast, are more relaxed and natural. Young de-emphasized the large driftwood log in the foreground and added two Indians, but his landscape remains relatively faithful to Möllhausen's view, which is quite similar to the actual topography (see fig. 6c). Both versions, however, bring the distant peak closer than it actually appears from the river.

6D
John James Young after F.W. von Egloffstein, *Cane Brake Cañon, From near Camp 16,* toned lithograph, from Ives, *Report* (1861), part 1, opp. p. 41, Amon Carter Museum Library

6C
Chimney Peak, March 1990

The outline of Chimney Peak also appears in the center of the lithograph titled *Cane Brake Cañon* made from a sketch by Egloffstein (cat. 6d). The print gives a generally vague and inexact idea of the topography from a bit downriver, perhaps at present Ferguson Lake, and also understates the size of the steamboat in proportion to the river.[7]

---

1   According to Gudde, *California Place Names*, p. 260, this formation, located in Imperial County, California, was first mentioned by Padres Pedro Font and Francisco Garcés, members of Juan Bautista de Anza's second expedition, on December 4, 1775, as "La Campana" (the bell) and "Peñon de la Campana" (rock of the bell), respectively. See Coues, ed., *On the Trail of a Spanish Pioneer: The Diary and Itinerary of Francisco Garcés*, vol. 1, pp. 161-162. Möllhausen, *Reisen*, chapter 7, in his entry from Fort Yuma, noted that the Indians called Chimney Rock "A-melle-e-quette" and that from the fort, it looked like a single "column which tapers toward the top."

2   Ives, *Report*, part 1, p. 48.

3   Compare Möllhausen, *Reisen*, chapter 9, entry for January 15, 1858, describing Chimney Peak, with Egloffstein's map 1 and USGS 7.5' series "Picacho" and "Little Picacho" quadrangles.

4   A related work, a finished watercolor of the steamboat which the artist gave to the Berlin Geographical Society near the end of his life, was destroyed during World War II. See Berlin Geographical Society 1904 list no. 17, *Dampfboot Explorer, benutzt zur Erforschung der Schiffbarkeit des Colorado vom kalifornisch[en] Golf aus bis zum Black Canon hinauf* ("Steamboat Explorer, used to determine the navigability of the Colorado from the Gulf of California to the Black Canyon").

5   In *Reisen*, chapter 8, Möllhausen describes the boat, Mr. Carroll, the engineer, and Captain Robinson, the pilot. He notes that the space in front of Captain Robinson's tiller was the "realm" of Dr. Newberry and himself. Unless otherwise noted, Möllhausen's observations in this entry are drawn from chapter 8 of *Reisen*.

6   Ives, *Report*, part 1, pp. 51-52.

7   Compared with the canyons they would soon see along the river, Cane Brake Cañon could hardly be termed a canyon, and modern maps do not name it. During our visit to the area, we identified it and noted the presence of canes and an old abandoned mine in the vicinity.

CAT. NO. 7

Balduin Möllhausen
Chimney Rock, From the East
Watercolor and gouache on paper
8⅜ x 11 in. (21.3 x 27.9 cm)
signed l.r.: "Möllhausen"
l.l. on paper mount: "12."
l.c. on paper mount: "Chimney rock, from the east"
ACM 1988.1.12

Cat No. 7

The steamboat's slow progress and frequent stops for sandbars allowed Möllhausen time to sketch Chimney Peak from several angles. This watercolor, taken from the east, has a vantage point similar to Möllhausen's image of the *Explorer* passing Chimney Peak (cat. no. 6) and to Egloffstein's view of the peak from Cane Brake Cañon (see cat. 6d). For this closer view, Möllhausen enlarged the peak considerably beyond its proper scale to emphasize the formation. Only the bow of the steamboat protrudes into the watercolor, which shows two men in a skiff in the center foreground sounding and hauling the anchor out in order to winch over a sandbar. This image of one man rowing while another stands with the sounding pole may relate to a field sketch titled *Eighteen Inches! Rocky Bottom!* (cat. 7a).[1] The lithograph in *Reisen* (cat. 7b) was probably taken from a second, now-lost watercolor of the scene.[2]

The surviving watercolor demonstrates some of Möllhausen's developing skill with this media. He laid in some of the areas in the water and sky with transparent washes, occasionally allowing the raised texture of the medium-rough watercolor paper to show through. For other highlights he used opaque layers of gouache; opaque white, for example, defines some of the ripples in the water or some of the lighter blades of cane on the riverbank, although he generally reserved more opaque treatments for land areas or billowy clouds.

1 The sequence of the field sketch in the notebook suggests that Möllhausen drew it sometime between February 21 and 25. However, since sounding was such a frequent occupation on the expedition, any date between January 11 and March 21 could be possible.

2 The lithographer, Heinrich Leutemann, substituted deer or elk and aquatic birds on a sandbar for the skiff and the steamboat's bow. Apparently Möllhausen made another watercolor view, now lost, of Chimney Peak from the south, according to the 1904 list, no. 19, for the Berlin Geographical Society: "Chimney Rock, von Süden aus." Although unlikely because of its order on the list, it is also possible that this lost sketch was of Chimney Rock in Nebraska.

7A

Balduin Möllhausen, *Eighteen Inches! Rocky Bottom! (Möllhausen and another Man in a Boat, Sounding.) [Möllhausen und ein anderer Mann in einem Boot beim loten]*, graphite on paper, from Sketchbook 3, Möllhausen Family, Bleicherode, Germany, photograph courtesy of Dr. Friedrich Schegk, Munich

7B

After Heinrich Leutemann after Balduin Möllhausen, *Schornsteinfelsen oder Chimney Peak*, toned, electrotype engraving or transfer lithograph, from Möllhausen, *Reisen* (Leipzig: Otto Purfürst, 1861), vol. 1, opp. p. 174

CAT. NO. 8

Balduin Möllhausen
Chimney Rock, From the North
Watercolor and gouache on brown-toned paper
8⅝ x 11⅜ in. (22.0 x 28.9 cm)
signed l.r.: "Möllhausen"
l.l. on paper mount: "13"
l.c. on paper mount: "Chimney rock from the north"
ACM 1988.1.13

Cat. No. 8

One of Möllhausen's most evocative works is this moonlight view of Chimney Peak from the north. The artist subtly rendered the details of a camp scene over a dark ground by using opaque white and by scraping the paper to evoke fires and highlights. The dark prow of the steamboat *Explorer* juts into the picture at far left along the riverbank; a minimal amount of light reveals the anchor and small fieldpiece on the prow, the highlighted front of the boiler, and the smokestack. The members of the expedition have grouped themselves around five campfires in a way that suggests a certain social hierarchy. Near the boat, three men—perhaps laborers or crewmen—smoke pipes while a sentry with rifle and bayonet stands guard. At the center campfire sit three cloaked figures, perhaps representing the Indian guides. Farther in the distance, other figures sit around fires or stand guard. Around the campfire in the right foreground, next to a pair of pyramidal tents, sit five figures playing musical instruments: a violin, two flutes, and one or two guitars.

Möllhausen undoubtedly incorporated material for this watercolor from his field sketch of four musicians (cat. 8a), which shows Möllhausen himself at left, wearing a Turkish fez and propped on one elbow as he plays his guitar and sings.[1] Next to Möllhausen is engineer Carroll, identifiable by the cap he wore in a previous sketch, sitting on a folding camp stool and playing his flute. Baron Egloffstein, wearing a wide-brimmed hat, sits on another camp stool and concentrates on playing his flute, while the bearded geologist, Dr. Newberry, stands and fiddles next to the campfire. A few pencil strokes at right even suggest the side of the tent, which appears in the right of the watercolor. To create the watercolor, Möllhausen rearranged the figures and added a cloaked and bearded figure to the left of the group—perhaps representing Lieutenant Ives.

Möllhausen's verbal descriptions in *Reisen* particularly complement this watercolor, providing an intimate glimpse of the explorers' lives. Noting that Egloffstein, Carroll, Dr. Newberry, Lieutenant Ives, and he had made room for their simple musical instruments aboard the *Explorer*, to "later provide us many a cheerful evening in the desolate wilderness,"[2] he described an evening of music on January 14, as they camped near Chimney Peak (camp no. 15):

> [U]ntil far into the night we sat together and leisurely played tunes on our instruments. For us there was a peculiar attraction to this activity and we devoted much enthusiasm to it. We

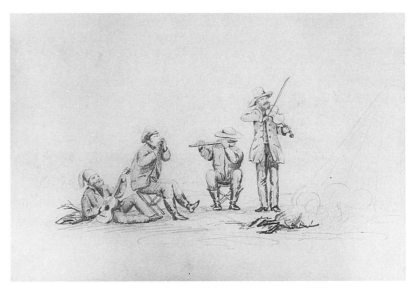

8A
Balduin Möllhausen, *Möllhausen, Carroll, Egloffstein, and Newberry Playing Musical Instruments* [*Vier Musiker*], graphite on paper, from Sketchbook 3, Möllhausen Family, Bleicherode, Germany, photograph courtesy of Dr. Friedrich Schegk, Munich

were the first to bring popular music into this wilderness, and the first to select the silent desert and the secluded river as audiences for our cheerful outpourings. Familiar sounds in the homeland are beautiful, but in a distant, foreign country, they force their way into the heart. Each chord touches a long-remembered string of memories. Even our rough soldiers appeared to be touched by this music in such surroundings. Many stayed nearby, for when the flames of our campfire shot up, they shone on more than one disheveled bearded figure behind us, stretched on the withered grass, listening to the music.[3]

The proximity of the peak to the south suggests that the sketch was made where the explorers camped on January 16 (camp 17), on the Arizona side of the river near present-day Adobe Lake, within the boundaries of the Imperial National Wildlife Refuge.[4] No prints reproduce the watercolor sketch exactly, but a remotely similar printed view of the peak taken from the north (cat. 8b) illustrates Dr. Newberry's geological report. The engraver of the print probably distorted the scale by shrinking the steamboat or enlarging the double pinnacles, which also appear too far apart.[5]

The explorers could most clearly distinguish the double pinnacles of Chimney Peak from the north. Dr. Newberry correctly guessed that it was "probably the northwestern prolongation of the middle range of the Purple Hills" (Chocolate Mountains) and that:

[l]ike the other peaks of the range it is composed of trap, and affords a striking example of the tendency to form columnar summits exhibited by all of the mountains of this vicinity. . . .by the erosion of rains and flowing streams, perpendicular walls are formed, and large masses usually exhibit mural faces. This will account for the peculiar outline [of] many of the trappean summits. . . Their great altitude, as compared with the mass of the ranges which they crown, is doubtless due to the resistance offered by their material to the atmospheric influences.[6]

1   From a later entry in *Reisen* we learn that the fez was a gift "from a dear friend, a famous Egyptian traveler." Toward the end of the expedition, Möllhausen, when hard-pressed to come up with a gift in return for the hospitality offered him by Zuni Governor Pedro Pino (see fig. 66), gave the latter this fez; *Reisen*, chapter 29.
2   Möllhausen, *Reisen*, chapter 8.
3   Möllhausen, *Reisen*, chapter 9.
4   Compare *Reisen*, chapter 9, and Egloffstein's map with USGS 7.5' series "Picacho SW" and "Picacho Peak" quadrangles. Picacho Peak is due south from Adobe Lake.
5   This same print also appeared in Alpheus Hyatt, "The Chasms of the Colorado," *American Naturalist* 2 (September 1868), opp. p. 360, plate 7.
6   Newberry, in Ives, *Report*, part 3, pp. 21-22. See also fig. 51.

8B
John James Young(?) after F. W. von Egloffstein or J. S. Newberry(?), *Chimney Peak from the North*, engraving, from Ives, *Report* (1861), part 3, p. 22, Amon Carter Museum Library

CAT. NO. 9

Balduin Möllhausen
Red Rock Gate
Watercolor and gouache on paper
7⅝ x 11 in. (19.4 x 27.9 cm)
Signed l.r.: "Möllhausen"
l.l. on paper mount: "21[Scratched out] 5"
l.c. on paper mount: "Red Rock gate."
l.r. on paper mount: "Red Rock Gate."
ACM 1988.1.5

On the cold morning of January 17, the *Explorer* found a good channel as it worked its way upstream between unusual rock formations. According to Möllhausen:

> At first black masses of basalt towered upward over the cotton-woods and willows on the banks to a modest height; but they seemed to grow in such dimensions as we moved north that they soon rose about one hundred and fifty feet high over the river's surface. The black stone was displaced again by red rock masses, chiefly porphyry which rose straight up out of the water 130 feet high. We named this point "Red Rock Gate." However, the rocks did not form connecting walls as the name "Red Rock Gate" implies. Instead, impressive groups stood across from each other separated by short spaces, and the foaming river wound between them.[1]

Cat. No. 9

The towers of Red Rock Gate straddle the Colorado River at the Imperial National Wildlife Refuge of Arizona and the Picacho State Recreation Area of California.[2] Möllhausen's watercolor records the actual topography fairly accurately when compared with a recent photo-graph (cat. 9b); he compressed the scene somewhat by narrowing the width of the gate, but not nearly as much as the engraver did in the print for Ives' *Report* (cat. 9a). The print for the *Report* was one of two such plates (the other being that of Bill Williams Mountain) for which a corrected proof and mockup still exists in the files of the National Archives. The War

9A
John James Young after Balduin Möllhausen, *Red Rock Gate*, engraving, from Ives, *Report* (1861), part 1, p. 50, Amon Carter Museum Library

9B
Red Rock Gate, March 1990

Department's Office of Explorations and Surveys apparently sent the earlier proofs back to the engraver for modifications because they were "not considered satisfactory."[3] In the corrected versions, the printmaker successfully translated the transparent tonalities of Möllhausen's watercolor rock formations into the linear medium of engraving and, despite the usual topographical distortions and exaggerations of scale, achieved a sense of the weight and mass of the actual geological structures.

In addition to typical details such as snags in the river and a flight of birds in V-formation overhead, both the watercolor and the print show an Indian raft in the left foreground. Since neither Möllhausen nor Ives referred to a raft at this point in their narratives, the artist probably included it in this composition merely to add a sense of scale. Möllhausen did sketch such a raft (cat. 9c) in 1854 for the Whipple report, which noted that Mojave rafts "were of simple construction, being merely bundles of rushes placed side by side, and securely bound together with willow twigs. But they were light and manageable, and their owners paddled them with considerable dexterity."[4] On February 1, 1858, in Monument Canyon, Lieutenant Ives and his men would meet "two Chemehuevis, with their wives, children, and household effects" on such rafts, and both Ives and Möllhausen described how their steamboat impressed the nautical natives. The white explorers could not help boastfully contrasting their superior technology with the primitive methods of the natives, although Möllhausen at least acknowledged the superiority of nature to both peoples when he wrote:

> They cowered on a raft fashioned of reeds and floated leisurely downstream. These poor people looked at our "firespewing canoe" with an expression of great fear. We looked at them and thought about how fate had created peculiar contrasts. On one hand, navigation in its infancy, the people and the vessel subject to the will of the elements. On the other hand, navigation in its highest state of perfection; obedient tools in the hands of mortals able to combat successfully against the elements. But both are parts of a grand sublime nature. It only needs to rain lightly, and primitive people as well as the vain disciples of civilization with all their works and inventions, would both disappear without a trace.[5]

1   Möllhausen, *Reisen*, chapter 9. Ives described the formation briefly in his *Report*, part 1, p. 50, and Newberry at greater length in the *Report*, part 3, p. 24.

2   Compare plat 6 of Egloffstein's original map; Egloffstein's printed map; USGS 1:250,000 series "Salton Sea"; 15' series "Picacho" quadrangle (1951); 7.5' series "Picacho SW" quadrangle. In his description in *Reisen*, Möllhausen mistakenly placed Red Rock Gate north of Sleeper's Bend, rather than vice-versa. David Miller has corrected this in his translation.

3   Letter, Captain A. A. Humphreys, Topographical Engineers, War Department, Office of Explorations & Surveys, to Head Engraver, Superintendent of Public Printing, March 23, 1861, in National Archives, RG 48, Office of Explorations & Surveys, "Correspondence Concerning the Colorado & San Juan Expedition, 1861," Box 1.

4   Whipple, "Report," in *Explorations and Surveys*, vol. 3, part 1, p. 117. The engraving after Möllhausen's drawing is on this page.

5   Möllhausen, *Reisen*, chapter 12. See also Ives, *Report*, part 1, p. 59.

9c
Balduin Möllhausen, *Mojave Raft*, pen and ink with wash and graphite on paper mounted on brown paper, Whipple Collection, Oklahoma Historical Society, photo 9256-ODU10

CAT. NO. 10

Balduin Möllhausen
Sleeper's Bend [Sleeper]
Watercolor and gouache on paper
8¼ x 10¼ in. (21.0 x 26.0 cm)
signed l.r.: "Möllhausen"
l.c. on paper mount: "Sleeper"
l.l. on paper mount: "6"
ACM 1988.1.6

Cat. No. 10

In his *Report*, Ives described a cluster of hills at Sleeper's Bend: "While turning a bend, a little while after passing the gate, we suddenly noticed upon the summit of a little hill on the left bank a ludicrous resemblance to a sleeping figure. The outlines and proportions were startlingly faithful, and the following sketch, hurriedly taken as the steamer passed, scarcely gives a true idea, and certainly not an exaggerated one, of the . . . likeness which presented itself from different positions for nearly a mile."[1]

Möllhausen was characteristically more excited: "We could not keep from shouting in astonishment when we saw it," he wrote in his diary for January 17, describing

> a horrible figure of a giant deep in sleep, reclining on a rocky bed, his head and back leaning in a comfortable manner. His head appeared to be covered with wooly hair. His eyes were tightly closed, and his chin tilted comfortably onto his deep chest. He lay on his back with his hands folded on his chest. His knees were raised somewhat and the tips of his toes pointed upwards. A wide garment or a blanket appeared to cover the lazy sleeper, but even through the garment one could see his gigantic but normal physique. So the giant lay there and slept. He had slept like this for thousands and thousands of years, and so he will rest until a mighty power some day shatters him.[2]

Möllhausen melodramatically speculated at length whether it was "the image of a civilization still slumbering on the Colorado" or "an image of the original inhabitants of the land . . . who as an entire nation will be hurled into his arms to sleep forever?" Echoing the vanishing race theory prevalent among his contemporaries, he sadly concluded that Nature had "created in the eternally sleep-

ing giant a monument which signifies the impending and immediate decline of a race."

The Sleeper, also visible in the distant right of Möllhausen's watercolor of Red Rock Gate (cat. no. 9), is in the foothills of the Trigo Mountains in Arizona's Imperial Wildlife Refuge. The Colorado now flows farther west of the formation, however, and the illusion of a sleeping figure no longer startles the passing visitor. Perhaps this is why it is not indicated on recent maps.[3]

---

1   Ives, *Report*, part 1, p. 50.
2   Möllhausen, *Reisen*, chapter 9.
3   Compare USGS 7.5' series "Picacho SW Quadrangle" and 15' series "Picacho Quadrangle" (1951) and Egloffstein's map 1. Also see plat 6 of Egloffstein's original map and USGS 1:250,000 map "Salton Sea."

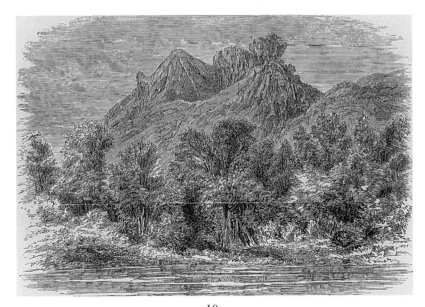

10A
John James Young after Balduin Möllhausen, *Sleeper's Bend*, engraving, from Ives, *Report* (1861), part 1, p. 51, Amon Carter Museum Library

CAT. NO. 11

Balduin Möllhausen
Lighthouse Rock
Watercolor and gouache on paper
7¾ x 11¼ in. (19.7 x 28.6 cm)
signed l.r.: "Möllhausen"
l.l. on paper mount: "No 14"
l.r. on paper mount: "Lighthouse Rock"
ACM 1988.1.14

Cat. No. 11

Around noon on January 17, the *Explorer* reached Lighthouse Rock not far above Sleeper's Bend. Lieutenant Ives noted that the circular pinnacle resembled a lighthouse blocking "the centre of the river, leaving a very narrow but fortunately unobstructed channel," and Dr. Newberry remarked on its composition of purple trachyte, "closely resembling the colored rocks of the Purple Hills."[1] Möllhausen likened it to "a giant sentinel splashed by foaming water" and noted that it was "almost 80 feet tall and exhibited a remarkably uniform shape like a sugar-cone." He speculated that Lighthouse Rock might have been attached at one time to the tall rocks on the left bank and remarked that it reminded him of the lighthouses he had "often seen on Lakes Erie and Michigan."[2]

The prepared watercolor sketch of Lighthouse Rock, seen from the south, exaggerates the size of the mountains in the distance but reproduces the pinnacle with some accuracy. Möllhausen added the snags and bighorn sheep at right partly for pictorial effect, but the expedition members actually did encounter them in this vicinity. He could have sketched some of the composition while they halted just before passing the rock; as they threw the gangplank down on the right bank and released surplus steam, he recorded that

> a herd of mountain goats, disturbed by the strange noise, ran
> up the steep slope of the nearby mountain on the other end of
> the large valley-shaped ravine. They hastily jumped from rock

11A
John James Young after Balduin Möllhausen, *Light-house Rock*, engraving, from Ives, *Report* (1861), part 1, p. 52, Amon Carter Museum Library

11B
Lighthouse Rock in March 1990

to rock, disappearing behind the next ridge. It was a beautiful sight. These heavily horned animals appeared to soar gracefully next to the abyss while hardly touching the ground with their agile hoofs. Some of our party followed them with rifles, but it would have been just as easy to have stopped a rolling stone as it would have been to make the frightened mountain goats wait for the pursuing hunters.[3]

Lieutenant Ives noted few signs of life but corroborated that "a dozen mountain sheep ('big horns') were seen scampering over a gravel hill near Light-house Rock, but not within shot from the bank of the river."[4] The engraver, closely adhering to Möllhausen's watercolor for the print in Ives' *Report*, included the foreground snags and the sheep (cat. 11a).

Captain Robinson skillfully steered the boat through the narrow channel at the right of the picture, then navigated a short, hazardous pass through the Chocolate Mountains. The navigational difficulties of this area were underscored a few weeks later, when Captain Johnson and the rival steamboat *Jesup* struck a rock and sank near Lighthouse Rock.

The name Lighthouse Rock remains on modern maps. It is located just south of Draper Lake in the Imperial Wildlife Refuge, La Paz County, Arizona (cat. 11b). The channel on the Arizona side of the Lighthouse is now often dry.[5]

1 Ives, *Report*, part 1, p. 50, and Newberry, in *Report*, part 3, p. 24.
2 Möllhausen, *Reisen*, chapter 9.
3 Ibid.
4 Ives, *Report*, part 1, p. 52. The Neues Palais Collection in Potsdam has a fine finished watercolor by Möllhausen of a mountain sheep.
5 Compare Ives' original and Egloffstein's printed map 1, and the USGS map 15' series "Picacho Quadrangle" (1951). See also Barnes, *Arizona Place Names*, p. 246.

CAT. NO. 12

Balduin Möllhausen
Mohave Indians
Watercolor and gouache on paper
8½ x 10⅜ in. (21.6 x 26.4 cm)
signed l.r.: "Möllhausen"
l.l. on paper mount: "19."
l.c. on paper mount: "Mohave Ind"
verso, center on paper mount: "Mohave Ind"
ACM 1988.1.19

Cat. No. 12

This watercolor relates to an incident on January 22 in the Palo Verde Valley near present-day Blythe, California.[1] Although Ives did not mention it, and no prints were made from the watercolor, Möllhausen recorded encountering Yuma Indians from a nearby village for several miles above Lighthouse Rock and, still farther north, both Chemehuevis and Mohaves in great numbers on the right bank.

Möllhausen wrote about the Mohave encounter in considerable detail, admiring the exuberant behavior of the native men and women as they hurried along the sandy bank to keep pace with the slowly moving steamboat:

The men stationed themselves on a protruding point on the bank and let the steamboat pass by. Then they hurried to the next prominence in order to enjoy the spectacle once again. By contrast, the women–actually girls from twelve to eighteen years of age–kept pace with the steamer and reached out toward us with pleading expressions which we found touching. Unable to resist, we threw beads and small swatches of cloth to them. It was a charming sight, these beautiful, voluptuous figures with their thick bark skirts reaching to the knee, their long, thick black hair hanging loosely around their painted faces, and their dark fiery eyes–a highly refined beauty which some who are scornful of the colored races might envy. They bounded through the deep sand at a hectic pace. One of them would tumble to the ground, dragging several down with her, but they would get up again as quick as lightning and try to overtake us. Then, when a string of beads fell between them on the ground, a short battle ensued. One could see a thick tangle of copper-colored arms and legs, fluttering strips of bark and whirling sand. But in the next moment, the frantic girls all disentangled and gracefully bounded along, laughing and rejoicing while begging for new presents. The Indian men and boys sneered in a characteristic way at the behavior of the female members of their tribe. They were obviously amused as the women tumbled over one another, but they appeared to consider it beneath their dignity to mingle with them or to beg strangers for presents in such a rambunctious way.[2]

In Möllhausen's watercolor, four girls with bark skirts and painted and tattooed faces, bodies, and limbs scramble toward the steamboat *Explorer*, suggested by just a portion of its red paddle wheel at far right.[3] A bearded white man (probably a Möllhausen self-portrait) sits on the wheelhouse, whistling at them and holding out a beaded necklace. Meanwhile, four of the more reserved native men, themselves painted and tattooed, wearing breechclouts and holding poles or lances, enjoy the spectacle.

According to Möllhausen, the natives finally reached an old deep riverbed which prevented them from following any further:

In the sharp angle formed by the two streams, the brown beauties stood there in water up to their knees and reached out their voluptuous arms to us. But the uncompassionate *Explorer* surged on, as if she wanted to remove us by force from the dangerous vicinity of these attractive Indian maidens, whose dark eyes look at us with a certain charm, despite the blue and red pigment which encircled them. We beckoned at the pretty daughters of the wilderness to follow us. They attempted in the same way to persuade us to turn around, and as we moved farther away from them, they jumped back onto the bank pouting, grabbed handfuls of sand with their little stubby hands, and amid the wildest laughter tossed it toward us. This was harmless Indian teasing. We laughed hard about it, and so did the warriors on the bank.[4]

The watercolor and the account contain many fascinating and evidently accurate ethnographic observations, including some that

Möllhausen had already made while with Whipple's expedition in 1854. Like the other lower Colorado River tribes, the Mohave and Chemehuevi wore their hair cut short over the eyebrows, but did not braid it,[5] and the man standing with his elbow on his companion's shoulder seems to have matted his hair up with a mixture of clay to destroy vermin.[6] Tattoos on the chin and face and body painting were common among Mohaves of both sexes.[7] Women and girls often wore only bark skirts, which Möllhausen thought presented "their voluptuous figures in an advantageous light"; from a distance, he noted, the women "looked very much like our ballet dancers."[8] The watercolor also shows two warriors wearing wristbands and various other jewelry.

1 Location based on comparison of narrative with plats 15, 16 and 17 of Egloffstein's original map, the printed map, USGS 15' series map "Blythe Quadrangle," and USGS 1:250,000 map "Salton Sea" (1959; revised 1969).
2 Möllhausen, *Reisen*, chapter 10. Although Möllhausen's narrative indicates the presence of both Mohaves and Chemehuevis, his title excludes the second group. The territory on the right bank around Blythe was generally Chemehuevi country. Evidently identification could be difficult, since sometimes the greatest differences between the two peoples were linguistic. For more information on Mohaves, see Weber, *Richard H. Kern*, pp. 177-179, and Wallace, "Sitgreaves Expedition," pp. 354-355.
3 In his *Report*, part 1, p. 66, Ives noted that the Mohaves were particularly interested in the "stern wheel. . . . It is painted red, their favorite color, and why it should turn around without any one touching it is evidently the theme of constant wonder and speculation."
4 Möllhausen, *Reisen*, chapter 10.
5 Möllhausen, *Diary of a Journey*, vol. 2, p. 245. See Kenneth M. Stewart, "Mohave," in Sturtevant and Ortiz, eds., *Handbook of North American Indians*, vol. 10, pp. 55-70, and Isabel T. Kelly and Catherine S. Fowler, "Southern Paiute," in *Handbook*, vol. 11, pp. 368-397.
6 The clay was customarily left on the hair for two or three days, then thoroughly washed out, restoring the hair to its usual luster. See Ives, *Report*, part 1, p. 39, and A. W. Whipple et al., "Report upon the Indian Tribes," in *Explorations and Surveys*, vol. 3, part 3, pp. 33-34.
7 Möllhausen, *Diary of a Journey*, vol. 2, pp. 242, 245, 250, 259; Stewart, "Mohave," pp. 57, 58, 62, 63.
8. Möllhausen, *Diary of a Journey*, vol. 2, p. 245; Ives, *Report*, part 1, p. 66. Whipple's "Report upon the Indian Tribes," in *Explorations and Surveys*, vol. 3, part 3, p. 33, noted that a Mohave skirt consisted "of two distinct articles; the back . . . simply of a mass of strips of the inner bark of cottonwood, united to a string which passes around the hips, while the apron is of twisted cords made of vegetable fibres, in various colors, hanging loosely from the girdle, to which they are bound."

CAT. NO. 13

Balduin Möllhausen
Riverside Mountain
Watercolor and gouache on paper
8 x 11 in. (20.3 x 27.9 cm)
signed l.r.: "Möllhausen"
l.l. on paper mount: "No 15"
l.r. on paper mount: "River side Mountain"
ACM 1988.1.15

Cat. No. 13

From January 24-31, the men on the slow-moving steamboat could view the Riverside Mountains of California. Möllhausen first mentioned this prominent land feature in his diary entry for January 25, as the explorers approached from the south. At this time he could hardly distinguish the Riverside Mountains from the "Half-Way Range," but as the steamboat drew closer, the mountains "appeared to grow . . . their form and outline became more distinct," and the two ranges separated.[1] The men camped on the right bank almost at the base of them on the nights of January 25 and 26 (camps 26 and 27). A sloping gravel plain, extending from near the mountains toward the river, exposed bedrock that seriously impeded the steamer's progress, and the *Explorer*, passing around the eastern base and to the north of the mountains, made only six and three-quarters miles on January 25 and even less on each of the next few days. This gave Möllhausen plenty of time to sketch the scenery in this vicinity.[2]

The finished watercolor that he prepared from these sketches shows the mountains and, at left, the gravel plain and some of the exposed bedrock, with sandbars and snags hinting at the treacherous navigation. Cranes or herons take flight in the center foreground, in a sequence almost prefiguring the studies of animals in motion by photographer Eadweard Muybridge. Möllhausen's peculiar, schematic treatment of shade and reflection in the water detracts by its vertical regularity.

The lack of color in this picture is disconcerting, since Ives, Möllhausen, and Newberry all commented on the great variety of colors exhibited in these mountains. Ives noted that "each successive range of mountains . . . presents more striking varieties and combinations of color, imparting a strange and novel beauty to the barren rocks. As the rays of the setting sun fall upon the rugged face of the Riverside mountain, and illuminate its crevices and hollows, tints of purple, blue, brown, almond, and rose color are brought out in gorgeous relief, and contrast singularly with the dull monotonous gray of the desert."[3] Similarly, Dr. Newberry described this "striking contrast of colors" and wrote:

This mountain is entirely destitute of vegetation, and, when seen at the distance of several miles, the patches of purple, brown, blue, ash, cream, red, &c. form a picture which would scarcely be exaggerated if represented by the colored diagrams of the geological lecture-room.

When the mountain is illuminated by the rising or setting sun this variety of color produces a novel and pleasing effect, redeeming, in some degree, the scene from the aspect of sterility and desolation which it would otherwise bear.[4]

13A
John James Young after Balduin Möllhausen, *Riverside Mountains*, engraving, from Ives, *Report* (1861), part 1, p. 54, Amon Carter Museum Library

Newberry speculated that the mountains could contain "rich mineral veins" of gold, silver, lead, iron, and copper, and from this analysis, Ives concluded that "a careful search might develop ample stores of treasure, which the close proximity of water transportation would greatly enhance in value."[5] Their suppositions may have proven correct; modern maps show a number of mines in the area.[6]

The print for Ives' *Report* (cat. 13a) differs from the watercolor in notable ways. The wood engraver cropped the whole right quarter of the picture, deleting the snags and much of the vegetation. By focusing on the mountain at the expense of the rest of the terrain, the print makes the mountains appear much nearer and larger than they do in Möllhausen's watercolor—or in reality.

---

1 Möllhausen, *Reisen*, chapter 10. See Gudde, *California Place Names*, for Half-Way Range. The Riverside Mountains appear on Egloffstein's maps as well as modern maps. See, for example, USGS 1:250,000 map "Salton Sea" and USGS 15' series "Needles" (1956; revised 1969). The Riverside Mountains rise to the south and west of the Half-Way Range, today known as the Big Maria Mountains. Ironbluff Mountain, which appears on Egloffstein's map, is evidently part of this range.

2 Möllhausen, *Reisen*, chapters 10, 11. The course of the Colorado apparently ran somewhat closer to the mountains in 1858 than it does today.

3 Ives, *Report*, part 1, p. 57.

4 Newberry in Ives, *Report*, part 3, pp. 26-27.

5 Ibid, part 3, p. 26, and part 1, p. 57.

6 See USGS 7.5' series "Vidal Quadrangle."

Cat. No. 14

Balduin Möllhausen
Distant View of the Monument Range
Watercolor and gouache on paper
8 x 11¼ in. (20.3 x 28.6 cm)
signed l.r.: "Möllhausen "
l.l. on paper mount: "17"
l.c. on paper mount: "Distant view of the Monument range"
ACM 1988.1.17

Cat. No. 15

Balduin Möllhausen
Monument Mountains
Watercolor and gouache on paper
8¼ x 10⅝ in. (21.0 x 27.0 cm)
signed l.r.: "Möllhausen "
l.l. on paper mount: "18"
l.c. on paper mount: "Monument M."
verso, center, on paper mount: "Monument M"
ACM 1988.1.18

Möllhausen prepared two watercolors of Monument Peak, California, elevation 2,453 ft., on the present border of San Bernardino County and the Colorado River Indian Reservation. He first mentioned the Monument range in his entry for January 29, when he noted that the explorers found themselves in a broad valley enclosed by mountains:

on the right bank . . . a plateau-like rock mass with perpendicular walls formed the eastern end of a deeply notched chain and a short distance beyond it a slender obelisk-like rocky tower rose up to the same height. The obelisk was joined to the main mass at the base. The whole group had the appearance of a gigantic distillery. I had special reasons for proposing that this prominent point be named Distillery Rock, but it was rejected. Perhaps they thought I was making an allusion to the peculiar American preference for whiskey. Enough! They selected a more noble name. The beautiful rock I have just described is now called Monument Mountain.[1]

The following day he wrote that the Monument Mountain Range "conferred a peculiar charm to the entire landscape without altering its terrible desolation" and that "as far as we could tell from a distance, the formations themselves appeared to be completely volcanic in nature."[2] Ives described Monument Mountain as "a slender and perfectly symmetrical spire that furnishes a striking landmark, as it can be seen from a great way down the river in beautiful relief against the sky."[3]

Cat. No. 14

Möllhausen evidently made his first, distant view of the mountain (cat. no. 14) from the south, somewhere between the Riverside Mountains and present-day Parker, Arizona.[4] Monument Peak rises in the far right distance, while other foothills and formations appear in the center and to the left. Driftwood again clogs the river channel. An Indian family, probably from a nearby Chemehuevi or Mohave village, is gathered on a projection of the river's right bank, waiting for the *Explorer* to pass.

Cat. No. 15

The second watercolor (cat. no. 15) shows Monument Peak from the south or southeast. It is much nearer, and the original sketch was probably made from somewhere above Parker—perhaps from Corner Rock (near present-day Headgate Rock Dam) in Monument Canyon.[5] In this view, which became the basis for an engraving in Ives' *Report*, Möllhausen included two pelicans near the foreground snags. The engraver once again compressed the scene and distorted some of the formations, and the pelicans have become hardly recognizable.

1 Möllhausen, *Reisen*, chapter 11.
2 Ibid. Möllhausen recognized their similarity to the mountains that he had seen on the Whipple expedition, which the geologist of that expedition, Dr. Jules Marcou, had already described. For the descriptions by Whipple and Marcou, see *Explorations and Surveys*, vol. 3, part 1, pp. 107-109; part 3, p. 37; and part 4, pp. 160-161.
3 Ives, *Report*, part 1, p. 55. He made his observation from Beaver Island (camp no. 32) near Corner Rock on January 31.
4 The Monument Mountains are now called the Whipple Mountains, with Monument Peak a part of them. See USGS 1:250,000 map "Needles" and 7.5' series "Whipple Wash" quadrangles. Cat. no. 14 is comparable to but not the same as the engraving in Ives, *Report*, part 3, p. 28. The Monument Range is on Egloffstein's map 1.
5 See USGS 1:250,000 series "Needles Quadrangle."

15A
John James Young after Balduin Möllhausen,
Monument Mountains, engraving, from Ives, *Report* (1861),
part 1, p. 55, Amon Carter Museum Library

Cat. No. 16

Balduin Möllhausen
Corner Rock
Watercolor and gouache on paper
8 x 10¼ in. irreg. (20.3 x 26.0 cm), torn
l.l. on paper mount: "v.23"
l.r. on paper mount: "Corner Rock"
ACM 1988.1.22

Cat. No. 16

Although the watercolor of Corner Rock has been damaged and only a fragment of the original remains, the print in Lieutenant Ives' *Report* records the full composition. Comparison of the fragment with its corresponding part in the print suggests that the engraver faithfully reproduced the watercolor, even including details from the original such as the snagged tree branches in the left foreground and two birds in the sky (cat. 16a).

"A few miles above" camp no. 32 at Beaver Island, where they stayed the night of January 31, Lieutenant Ives wrote, "the river wound around the base of a massive rock, into which a deep groove had been cut by the ceaseless flow of the stream. This point may be considered the southern entrance of the cañon through the Monument mountains."[1] Möllhausen recorded on February 1 that this "rocky projection" on the left bank

> rose to a height of about thirty feet and extended far out into the river. The river formed an angle there, so we conferred the name Corner Rock on this point. The rock itself was attached to the gravel plain, or rather appeared to be an extension of it. As we traveled close past it, we could clearly see two layers of conglomerate upon which layers of gravel rested. The upper edge was composed of a thick layer of basalt, and like the underlying strata, it dipped slightly from west to east.[2]

Corner Rock may be identical to modern Headgate Rock in San Bernardino County, California, near which Headgate Rock Dam now stands just above the town of Parker, Arizona.[3]

---

1 Ives, *Report*, part 1, p. 57.
2 Möllhausen, *Reisen*, chapter 12. Dr. Newberry also described Corner Rock in Ives, *Report*, part 3, p. 27.
3 Compare Egloffstein's maps and USGS 1:250,000 map "Needles" and 7.5' series "Parker" quadrangles.

16A
John James Young after Balduin Möllhausen, *Corner Rock*, engraving, from Ives, *Report* (1861), part 1, p. 57, Amon Carter Museum Library

CAT. No. 17

Balduin Möllhausen
Monument Cañon (Cañon through the Monument)
Watercolor and gouache on paper
7⅝ x 10 in. (19.4 x 25.4 cm)
signed l.r.: "Möllhausen"
l.c. on paper mount: "Cañon through the monument"
verso, center on paper mount: "Canon thro Mountain range"
ACM 1988.1.2

Cat. No. 17

Monument Canyon extends along the Colorado for about ten miles, from Corner Rock northeast to the mouth of Bill Williams River. Möllhausen apparently sketched his original view somewhere in the southern part of the canyon, looking toward the northeast as they passed through on February 1. The watercolor may show what are today known as the Buckskin Mountains and "the Mesa" in the distance.[1]

Lieutenant Ives described their entrance into the canyon:

Immediately above [Corner Rock] the river grew narrower and deeper, and the hills crowded closely upon the water's edge. The regular slopes gradually gave place to rough and confused masses of rock, and the scenery at every instant became wilder and more romantic. New and surprising effects of coloring added to the beauty of the vista. In the foreground, light and delicate tints predominated, and broad surfaces of lilac, pearl color, pink, and white, contrasted strongly with the sombre masses piled up behind. In their very midst a pile of a vivid blood red rose in isolated prominence. A few miles higher a narrow gateway opened into the heart of the mountains. On one side of the entrance was a dark red column, on the other a leaning tower of the same color overhung the pass, the ponderous rock seemingly ready to fall as we passed beneath.[2]

Möllhausen observed that initially the canyon walls were "not very significant, but increased gradually in height and extent as we moved upstream. Finally the magnificent formations surrounded

17A
John James Young after Balduin Möllhausen, *Monument Cañon*, engraving, from Ives, *Report* (1861), part 1, p. 57, Amon Carter Museum Library

us on all sides," although "the current was too rapid for us to fully enjoy the beautiful scenery."[3] Dr. Newberry also noted the canyon's "wild and picturesque scenery, the effect of the varied outline of its walls being heightened by the vivid and strongly contrasted colors which they exhibit."[4]

Perhaps to emphasize the navigability of the river, the snags that appear so prominently in Möllhausen's watercolor are not in the print for Ives' *Report* (cat. 17a). The engraver also brought the canyon walls closer to the picture plane, creating a grander and more imposing impression of their relative height.

---

1 Compare Ives' map with USGS 1:250,000 map "Needles" and 7.5' series "Cross Roads," "Gene Wash," "Monkeys Head," and "Osborne Well" quadrangles. We could not identify the site for this watercolor during a short visit to the Parker Dam area, but we did not travel on the river all the way down to Headgate Rock Dam.

2 Ives, *Report*, part 1, pp. 57-58.

3 Möllhausen, *Reisen*, chapter 12.

4 Newberry, in Ives, *Report*, part 3, pp. 27-28.

Cat. No. 18

Balduin Möllhausen
***Left Bank, Four Miles South of Bill Williams Fork
(Shore on Lower Colorado)***
Watercolor and gouache on paper
8¼ x 11½ in. (21.0 x 29.2 cm)
signed l.r.: "Möllhausen"
l.l. on paper mount: "22"
l.c. on paper mount: "L. Bank, 4 m. south of Bill Williams fork."
ACM 1988.1.21

Cat. No. 18

Describing the section of Monument Canyon just south of Bill Williams' River, Möllhausen wrote: "we caught sight of wonderful formations, which were all the more surprising, since they were composed of solid volcanic rock, which resistant to the elements had retained the unaltered character of its original formation." Likening the formations to "a castle with uniform architecture and ramparts and walls," terraces, watchtowers, and angular walls "projected up high," he noted that "these natural edifices were connected to one another by long walls which again were richly adorned with pillars and small towers." Strata of basalt and lava ran in "parallel layers for a great distance . . . [and] stood out clearly on the rugged lateral walls, sometimes in layers, sometimes resembling columns; because of this, the remarkable formations looked even more like elaborate buildings. There was something wildly romantic about our entire surroundings," and "the cacti scattered on the slopes stood like isolated sentinels or a mighty candelabrum, [and] tended to heighten rather than lessen the lifeless quality of the scene."[1]

Lieutenant Ives described "rich hues of blue, green, and purple, relieved here and there by veins of pink and white, . . . blended in brilliant confusion upon the sides of the cañon, producing a weird-like and unearthly effect, which the fantastic shapes and outlines of the enclosing walls did not diminish." Probably thinking particularly of this part of Monument Canyon, Ives called it a

18A
John James Young after Balduin Möllhausen, *Shore on Lower Colorado*, toned lithograph (black with ruled gray), from Ives, *Report* (1861), part 3, p. 20, Amon Carter Museum Library

18B
S. S. after Heinrich Leutemann after Balduin Möllhausen, *Felsformation in der Nähe der Mundung von Bill Williams fork*, toned electrotype engraving or transfer lithograph, from Möllhausen, *Reisen* (1861), vol. 1, opp. p. 238

18c
Balduin Möllhausen,
Cereus Giganteus, graphite with ink
wash heightened with opaque
white on paper, National Archives,
Washington, D.C.

18d
Balduin Möllhausen,
Cereus Giganteus, graphite with ink
wash on paper, National Archives,
Washington, D.C.

18e
Monument Canyon, March 1990

"fairy-like pass, where every turn varied and heightened the interest of the pageant."[2]

The watercolor and prints of this section of Monument Canyon (cats. 18a and 18b) depict an area currently occupied by the Holiday Harbor Marina and Beach Club in northwest La Paz County, Arizona. A comparison of the actual formations, including Castle Rock in Buckskin Mountains State Park, with those in the watercolor reveals some distortion and alteration by the artist; the landforms are compressed in order to include them within the picture.[3] The steamer evidently passed this section of the river at a fast pace on February 1, which may have given Möllhausen less time than normal to sketch these formations. He records that they had worked up a full head of steam, and the expedition certainly made better mileage than on the last days in January.[4]

The saguaro cacti that the artist mentioned also dot the landscape of his watercolor. Earlier, while with the Whipple expedition along Bill Williams River, Möllhausen had made at least two detailed sketches of such cacti (cats. 18c and 18d), which may have served as models for an engraved illustration in his *Diary of a Journey*.[5]

Both the cacti and the Indians in the watercolor titled *Left Bank, Four Miles South of Bill Williams Fork* are out of scale with the terrain, appearing somewhat larger than they would in reality (cat. 18e). Möllhausen probably added the Indians for pictorial effect, since none of the accounts mention an encounter with them here. Despite this artistic license, Möllhausen does capture in pictorial form the "isolation," "desolation," and "wildly romantic," "weird-like and unearthly effects" that both he and Ives described.

John James Young or the lithographer reinterpreted some of the shapes in Möllhausen's watercolor for the lithograph in Ives' *Report* (cat. 18a). The receding perspective of some strata in the watercolor have disappeared in the print, and the shapes of the giant cacti are distorted. The toned lithographic print for the German edition of Möllhausen's *Reisen* rearranges the foreground (cat. 18b). Heinrich Leutemann, the animal illustrator who re-worked Möllhausen's sketches for this version, omits the Indians and introduces herons and bighorn sheep.[6]

---

1 Möllhausen, *Reisen*, chapter 12.

2 Ives, *Report*, part 1, pp. 57-58.

3 The formation at left in the picture may be Monkey's Head, elevation 1587; the one at right may be Castle Rock, elevation 1750; and at distant right may be Giers Mountain, elevation 1888 ft., or a peak near it. See USGS 1:250,000 map "Needles" and 7.5' series "Monkeys Head," "Osborne Well," "Cross Roads," and "Gene Wash" quadrangles. Granger, *Arizona's Names*, p. 125, says Castle Rock "is also known as Mohave or Pulpit Rock. According to the informant [E. R. Householder, who in the early twentieth century compiled an official map of Mohave County], when the Colorado River used to be here, 'The rock stuck up at low stages and looked like a castle.'"

4 Möllhausen, *Reisen*, chapter 12.

5 When we visited the area just south of Parker Dam and the mouth of Bill Williams River in 1990, we could see no saguaros along the banks. We wondered whether the encroachment of recent developments had brought about this disappearance. Möllhausen described such cacti at length in *Diary of a Journey*, vol. 2, pp. 218-221, 230. The Whipple Report, in *Explorations and Surveys*, vol. 3, part 1, p. 101, has an engraving of a *Cereus Giganteus, on Bill Williams' Fork*, based on a sketch by John James Young, which is today in the Oklahoma Historical Society. Young in turn copied his sketch from Lieutenant John C. Tidball, whom he acknowledged in the inscription below the image. Tidball, like Möllhausen, was a member of the Whipple expedition.

6 The toned lithograph titled *Klippformation nära Bill Williams fork mynning*, in the Swedish version of the diary, *Resor* (Stockholm: 1867), is quite similar to the print in the German edition (cat. 18b).

CAT. NO. 19

Balduin Möllhausen
Four Miles North of Bill Williams Fork, View Down the River (Monument Range from the North)
Watercolor and gouache on paper
7⅝ x 11 in. (19.4 x 27.9 cm)
signed l.r.: "Möllhausen"
l.l. on paper mount: "32"
l.c. on paper mount: "4 m. north of Bill Williams Fork view down the river."
ACM 1988.1.30

Cat. No. 19

Having passed through the northern entrance of Monument Canyon at the site of present-day Parker Dam, the expedition emerged "into a comparatively open valley," where some expedition members who had been with Lieutenant Whipple in 1854 recognized the mouth of "Bill Williams' Fork." Whipple and his men had struck the Colorado River here on February 20, 1854, after following Bill Williams' River for many miles. At that time Möllhausen had written: "we did not imagine, on the 19th of February, that the chain of rocks before us lay on the western shore of the Colorado, so that when (on the 20th) a sudden turn of the valley brought us in full view of the broad majestic river, the sight was as unexpected as welcome."[1] However, when they reached the same point with Lieutenant Ives, some of the men were skeptical at first, since they only found "a very narrow gulley, through which a feeble stream was trickling," whereas in February of 1854 the tributary had been "about thirty feet wide." Surmising that there had been a two- or three-year drought, Ives reported: "The former mouth is now filled up, and overgrown with thickets of willow. . . . The Colorado, according to the Indians, is as low, proportionally, as its tributary."[2] Möllhausen noted that "only prominent mountain formations which I had carefully sketched" in 1854 (cat. 19c) finally convinced them that they had reached the same spot.[3]

The artist did not record sketching the mouth of Bill Williams' River again while on the Ives expedition, but he clearly had the opportunity.[4] In *Reisen* he described a hunting excursion up the fork in the late afternoon of February 1, 1858. As he looked around, he noticed familiar rock masses which "stood unchanged in the distance" and recalled that "once I attempted to portray their jagged lines accurately on paper." The next day, after the steamer had proceeded a little farther up the river, Möllhausen, Dr. Newberry, and several others returned to the same location to fish. The watercolor *Four Miles North of Bill Williams Fork, View Down the River* depicts an area near where the expedition spent the night of February 2 (camp 34). Stormy weather and the shallowness of the river had prevented much travel that day, and they camped uncomfortably in a small expanse "overgrown with chaparral."[5]

The watercolor served as the model for the engraving in Ives' *Report* titled *Monument Range from the North* (cat. 19a). Möllhausen's watercolor brought the formations closer to the viewer but kept their shapes and details generally accurate. The engraver compressed the scene even more, increased the height and steepness of the mountain slopes, and removed the driftwood and snags in the river.

A similar engraving, *View of Northern Entrance to Monument Cañon*, accompanied the geological report of Ives' expedition (cat. 19b). The original sketch for this engraving apparently no longer exists, but the engraving closely resembles the subject of Möllhausen's watercolor, showing the eastern end of the mesa seen in the lefthand portion of the watercolor. The engraver apparently increased the height and decreased the length of the mesa to make it fit the rectangular format of the picture, and either he or the artist of the original sketch apparently enlarged the conical hills at left, which Dr. Newberry had noted in his report.[6]

Although members of the expedition referred to the mountains at right and left in the picture as the Monument range, today the mesa at left is known simply as "The Mesa," part of the Buck-

19A
John James Young after Balduin Möllhausen, *Monument Range from the North*, engraving, from Ives, *Report* (1861), part 1, p. 60, Amon Carter Museum Library

19B
John James Young after J. S. Newberry, Balduin Möllhausen, or F. W. von Egloffstein, *View of Northern Entrance to Monument Cañon*, engraving, from Ives, *Report* (1861), part 3, p. 28, Amon Carter Museum Library

19c
Balduin Möllhausen, *Rio Colorado Below Junction of Bill Williams Fork*,
ink wash with opaque white on brown paper mounted on buff paper,
Whipple Collection, Oklahoma Historical Society, photo 9256-OD16

19d
Whitset intake on Lake Havasu, March 1990

skin Mountains, and the formations at right form part of the Whipple Mountains. The view today also includes Lake Havasu and the Whitset intake station for the Colorado River Aqueduct (cat. 19d).[7]

1 Möllhausen, *Diary of a Journey*, vol. 2, p. 239.
2 Ives, *Report*, part 1, pp. 58-59.
3 Möllhausen, *Reisen*, chapter 12. Möllhausen's 1854 sketch (cat. 19c) formerly belonged to the family of A. W. Whipple and is now in the Oklahoma Historical Society in Oklahoma City. Eduard Hildebrandt redrew this for the frontispiece of *Tagebuch einer Reise vom Mississippi nach der Küsten der Südsee* (1858) (see fig. 42), and a reduced version of the engraving appeared in the second volume of the English translation; see Möllhausen, *Diary of a Journey*, vol. 2, opp. p. 239.
  Before his third trip to America, Möllhausen also gave a finished watercolor of this subject, titled *Mündung von Bill Williams Fork in den Colorado*, to the King of Prussia; it is in the Neues Palais Collection, Potsdam, as is a fine watercolor of cliffs near Bill Williams' River (*Fels Formationen nahe von Bill Williams Fork, Mündung in den Colorado Grande*). This depicts the same area as the Oklahoma Historical Society's graphite sketch titled *Last Gate, Bill Williams' Fork* (see fig. 34), by Lieutenant John C. Tidball, who was also with the Whipple expedition, and the finely drawn graphite, ink, and wash copy of Tidball's sketch (see fig. 35) by John James Young for the engraving in Whipple's "Report" in *Explorations and Surveys*, vol. 3, part 1, p. 108. Möllhausen's and Tidball's views depict a natural gate formed where Bill Williams River cuts through the Aubrey Hills and "The Mesa" of the Buckskin Mountains near an overlook on present Arizona State Highway 95. See USGS 7.5′ series "Monkeys Head" quadrangle.
4 Three now-lost sketches that formerly belonged to the Geographical Society in Berlin also depicted this vicinity. They may have related to either expedition; a 1904 list describes them only as:
  no. 14. Bill Williams Fork
  no. 15. Mündung der Bill Williams Fork in den Colorado
       [Mouth of Bill Williams Fork in the Colorado]
  no. 24. Colorado Ufer . . . Bill Williams fork, v. N.
       [Colorado Riverbank . . . Bill Williams fork, from the North]
5 Möllhausen, *Reisen*, chapter 12.
6 See Newberry's observations in Ives, *Report*, part 3, p. 28.
7 See USGS 7.5′ series "Monkeys Head" and "Gene Wash" quadrangles.

CAT. NO. 20

Balduin Möllhausen
Mount Whipple and Boat Rock
Watercolor and gouache on paper
7⅝ x 10⅝ in. (19.4 x 27.0 cm)
signed l.r.: "Möllhausen"
l.l. on paper mount: "No 24"
l.r. on paper mount: "Mount Whipple & Boat Rock"
ACM 1988.1.23

20A
John James Young after Balduin Möllhausen, *Mount Whipple*, engraving,
from Ives, *Report* (1861), part 1, p. 61, Amon Carter Museum Library

Möllhausen recorded that on February 4: "An immense isolated rock caught my eye among the jumbled mountains . . . not far from the river. It was striking because of its height (approximately 1200 feet), and also because of its fine bold profile. Lieutenant Ives named it Mount Whipple in honor of Captain Whipple, our former joint commander. Farther upstream, another rock which jutted up out of the river formed a boat-shaped island; we named this rock 'Boat Rock.'"[1]

Cat. No. 20

Möllhausen distinguished these features in his watercolor, which the engraver faithfully copied for Ives' *Report* (cat. 20a). In the print, however, the light source is from a different direction and the vertical reflection in the center of the watercolor has been diffused.

The artist apparently made the sketch for the watercolor from a position between present-day Whipple Point or Whipple Bay on the San Bernardino County, California side and Beaver Point on the Mohave County, Arizona side of Lake Havasu, looking south-southeast. In general, the topographical outline is quite exact—except that Boat Rock is now underwater.[2]

---

1 Möllhausen, *Reisen*, chapter 12. In his *Report*, part 1, p. 60, Ives referred to Mount Whipple as "a high peak of the Monument range." Modern maps do not specifically name Mount Whipple but refer to the Monument Mountains as the "Whipple Mountains." Gudde, *California Place Names*, p. 388, notes: "the Von Leicht-Craven map shows the highest peak of this range as Mount Whipple," but the Geological Survey mapping in 1902-03 "applied the name to the great mountain mass in the bend of the Colorado, which includes Monument Range. . . . The old name is preserved in Monument Peak." These mapmakers could have benefitted from Möllhausen's watercolor and diary.

2 The site is only a few miles southeast of Lake Havasu City and its famous London Bridge. On our March 1990 boat visit to the site, we passed over the area in the middle of the channel where we guessed Boat Rock might be—where a sonar reading suddenly went from fifty to ten feet in depth. Compare USGS 1:250,000 map "Needles"; 17' series "Whipple Mountains"; 7.5' series "Lake Havasu City South," "Whipple Wash," and "Gene Wash" quadrangles; and Egloffstein's map 1. The Fish n' Map Co. "Lake Havasu" Map also designates "Steamboat Cove" just above Beaver Point.

CAT. NO. 21

Balduin Möllhausen
Distant View of the Mohave Range or Needles
Watercolor and gouache on paper
8 x 10¾ in. (20.3 x 27.3 cm)
signed l.r.: "Möllhausen"
l.l. on paper mount: "26"
l.c. on paper mount:
"Distant view of the Mohave range or Needles"
ACM 1988.1.25

Cat. No. 21

The Needles or the Mohave Range—a cluster of three slender and prominent pinnacles from which the present town of Needles, California, derived its name[1]—rise in the distance just left of center in this watercolor. The view, which was not reproduced in Ives' *Report* or Möllhausen's *Reisen*, is generally quite accurate from a topographical standpoint. It faces north, probably across what modern maps refer to as Blankenship Bend in Havasu National Wildlife Refuge.[2]

Möllhausen first saw this range while on the Whipple expedition in February 1854. In fact, Lieutenant Whipple had given the mountains their double names "Mohave" or "the Needles" at that time, and the lieutenant had even attempted to note their Quechan or Yuma name: "Ascientic-hâbî."[3] Travelling with Ives in 1858, Möllhausen first glimpsed the Needles again on February 4, after climbing some lava-like hills when the steamer stopped near Boat Rock to obtain wood. For the next several days, as they passed through the Chemehuevi Valley, they could see these "fantastically formed peaks" whenever the banks of the river were not too steep to prevent a distant view. In an entry for February 7, Möllhausen described the Needles at sunset, together with pelicans and cranes, and in fact two pelicans appear in the watercolor, skillfully rendered in opaque white along with their watery reflections.

The expedition evidently reached the site where Möllhausen made the original sketch for this watercolor on February 9. According to his narrative, "after a sudden bend in the river, the wonderfully formed towers and domes of the Needles lay immediately before us. The inviting southern edge of the Mohave Valley greeted us."[4] Snags and sandbars visible in the watercolor indicate treacherous obstacles to navigation, and Lieutenant Ives reported that "the channel was again tortuous . . . As we approached the mouth of the cañon through the Mojave mountains, a roaring noise ahead gave notice that we were coming to a rapid. . . . After ascending a few yards a harsh grating noise warned us that we were upon a rocky shoal."[5] All members of the expedition worried about what lay ahead, but some of their fears had been allayed just prior to reaching this bend, when they met a family of Mohaves floating downriver on a raft and interpreted this as a favorable sign that the steamboat might also negotiate the next few miles upriver.[6]

---

1 Gudde, *California Place Names*, p. 232.
2 See USGS 1:250,000 map "Needles" and compare with Egloffstein's printed map 1.
3 Möllhausen, *Diary of a Journey*, vol. 2, pp. 251-253, does not call the Needles by their name, but he was probably referring to them on February 23, 1854, when he wrote about "the sharp mountain ridge that we had seen before us in the morning was now gradually left behind, and the domes and peaks rose like regular towers and obelisks up into the clear air, in which the smallest line of those blue mountain masses was clearly distinguishable." Whipple's report in *Explorations and Surveys*, vol. 3, part 2, p. 30, explains the etymologies of names in a footnote that begins: "Mr. Blake has lately called my attention to the etymology of the word 'Mojave.' It appears to be formed of two Yuma words—hamook (three) and hâbî (mountains)—and designates the tribe of Indians which occupies a valley of the Colorado lying between three mountains."
4 Möllhausen, *Reisen*, chapter 12. Newberry noted that at the approach to Mojave Cañon and the Needles: "The foot-hills . . . exhibi[t] a variety of colors quite as vivid and as strongly contrasted as those of . . . Monument cañon. The colors are here, however, not precisely the same, the most conspicuous being bright red, green, and white of the stratified, and brown and purple of the erupted rocks." Newberry in Ives, *Report*, part 3, p. 30.
5 Ives, *Report*, part 1, p. 63.
6 Möllhausen, *Reisen*, chapter 12.

CAT. NO. 22

Entrance of Mojave Cañon (Mouth of Mohave Cañon)
Watercolor and gouache on paper
8¼ x 11 in. (21.0 x 27.9 cm)
signed l.r.: "Möllhausen"
l.l. on paper mount: "No 27"
l.c. on paper mount: "Entrance of Mojave Cañon"
ACM 1988.1.26

Cat. No. 22

Möllhausen's watercolor of the southern entrance to Mohave Canyon faithfully reproduces the topography of this site, a little over a mile from where he made his previous, distant view of the Needles. The dark-colored Mohave Rock, whose name remains on modern maps, may be seen just to the right of center, marking the entrance to the canyon, while the pinnacles of the Needles rise in the far distance.[1] The sand or gravel bar at left still impedes navigation, and brushstrokes of opaque white just below the entrance denote rapids. For Ives' *Report*, the engraver compressed the view into a narrower format by producing taller mountains and rearranging topographical features; the print also reinterprets the reflections in the water (cat. 22a).

Ives described the entrance to the canyon as "a low purple gateway and a splendid corridor, with massive red walls. . . . At the head of this avenue frowning mountains, piled one above the other, seemed to block the way," and Dr. Newberry noted that the gateway was "composed of trap and a massive and highly metamorphosed conglomerate, which varies in color from umber brown to blood red."[2] Rapids at the entrance delayed the steamboat's progress on February 9, and while Captain Robinson went up in a skiff to search for a passage, Möllhausen and the others enjoyed the "magnificent vistas which the picturesque scenery offered." However, the artist wrote that he "could not help feeling anxious as I looked over at the entrance and saw the stream suddenly

disappear behind overhanging volcanic rocks." He found it impossible "to guess the direction the broad waterway might lead. I sat on the small platform and sketched in quiet admiration."[3]

---

1   Compare USGS 1:250,000 map "Needles" and Egloffstein's map between camps 39 and 40.
2   Ives, *Report*, part 1, pp. 63-64, and Newberry in ibid., part 3, p. 30.
3   Möllhausen, *Reisen*, chapter 12.

22A
John James Young after Balduin Möllhausen, *Mouth of Mojave Cañon*,
engraving, from Ives, *Report* (1861), part 1, p. 63,
Amon Carter Museum Library

CAT. NO. 23

Balduin Möllhausen
Mohave Cañon 1
Watercolor and gouache on paper
8¼ x 11 in. (21.0 x 27.9 cm)
signed l.r.: "Möllhausen"
l.l. on paper mount: "28"
l.c. on paper mount: "Mohave cañon 1"
ACM 1988.1.27

Cat. No. 23

Möllhausen evidently made several sketches of the scenery within Mohave Canyon: one field sketch and one watercolor survive, and another sketch (and perhaps a fourth) are known through prints. The inscription "Mojave Cañon 1" suggests that the watercolor was the first in topographical sequence, and indeed it depicts an area just above the southern entrance of the canyon, probably looking northwest.[1] A portion of so-called Picture Rock is at right in the watercolor. In actuality, the rocks in the center and right of the picture exhibit natural "arched windows," which the artist described as the steamer passed by on February 9. These arches, placed at various angles, "opened and allowed the radiant western sky to penetrate through. Walls that moved along behind them created the distorted impressions of bridges. Apparent gates burst asunder and showed themselves to be two distant columns bending over toward one another at widely separated distances, or slid past one another in opposite directions."[2] Lieutenant Ives exclaimed that "a scene of such imposing grandeur . . . I have never before witnessed," and Dr. Newberry was equally impressed.[3]

The topography in John James Young's engraving titled *Mojave Cañon* (cat. 23a), crediting Möllhausen with the original view, is largely unrecognizable. The precariously balanced, angular cliffs are so exaggerated that they bear little resemblance to reality. The scale of the steamer in the center foreground is much too small and the wall of mountains in the distance much too large. Since Möllhausen's sketches are generally topographically accurate, it is tempting to blame the printmaker for these distortions, but the latter, with a little of his own imagination, was probably just following the awe-struck reports of the explorers. Ives wrote of "majestic cliffs, hundreds of feet in height," that "rose perpendicularly from the water" on either side of the river, while "[f]ar above, clear and distinct upon the narrow strip of sky, turrets, spires, jagged statue-like peaks and grotesque pinnacles overlooked the deep abyss."[4] Möllhausen described the canyon as a "gigantic corridor" in which

> immense layers of red sandstone were piled upon one another and rose straight up to a height of 500 feet. Since these long interconnected walls were lower and closer together toward the north, the gorge appeared twice as long as it actually was. Everything appeared larger than it was, for the reflection in the still water looked exactly like a continuation of the overhanging rocks, and since the cloudless sky bathed the middle of the stream in a transparent light blue, one could imagine himself suspended over an endless chasm.[5]

The river itself now reaches farther up the canyon walls, so modern visitors do not see such a dramatic effect.

23A
John James Young after Balduin Möllhausen, *Mojave Cañon*, steel engraving, from Ives, *Report* (1861), part 1, opp. p. 63, Amon Carter Museum Library

A field sketch inscribed "3te Canon, Mohave vall." (cat. 23b) possibly served as the model for the print, but the two are so different that another sketch may have been involved. The field sketch apparently faces northwest up the canyon, below Pulpit Rock, and is considerably more accurate than the print. While the rocks and cliffs are certainly impressive, they do not hang over the river as Möllhausen's narrative and the print suggest. A line of birds in the sketch, skimming over the water ahead, suggests the motion of the approaching steamer.

Having passed through most of the canyon on the afternoon of February 9, the explorers camped for the night near its northern entrance. From near this camp Möllhausen apparently made another sketch, which served as the model for the engraving titled *Head of Mojave Cañon* (cat. 23c). Möllhausen, rather than Egloffstein, probably made the original sketch, since the sequence of birds taking flight is similar to the motifs in Möllhausen's earlier field sketch (cat. 23b) and in his watercolor of Riverside Mountain

23c
John James Young after Balduin Möllhausen?, *Head of Mojave Cañon*, engraving, from Ives, *Report* (1861), part 1, p. 65, Amon Carter Museum Library

(scc cat. no. 13). Although the topography depicted in this print is somewhat distorted and exaggerated, it is clearly visible today (cats. 23d-e). Another channel of the river sometimes flows through the area in the center of the picture, but the river's main course is to the right, just beyond the rocks.

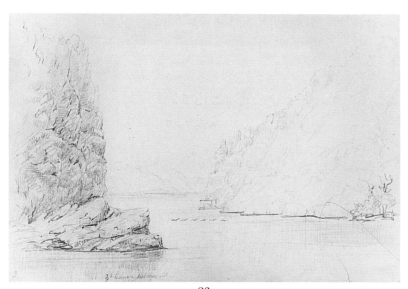

23b
Balduin Möllhausen, *Third Canyon, Mohave Valley*, graphite on paper, from Sketchbook 3, Möllhausen Family, Bleicherode, Germany, photograph courtesy of Dr. Friedrich Schegk, Munich

1   For the topography of this area, see USGS 7.5' series maps "Castle Rock" and "Topock" quadrangles, 15' series "Topock Quadrangle," Egloffstein's map 1, between camps 39 and 40; and plats 31 and 32 of Ives' original map.
2   Möllhausen, *Reisen*, chapter 13.
3   Ives, *Report*, part 1, p. 64. In his report in ibid., part 3, p. 30, Newberry wrote: "The cañon itself presented the most varied and interesting scenery we had met with, its walls being particularly ragged and picturesque in outline, and the colors such as to produce the most unusual and surprising effects."
4   Ives, *Report*, part 1, pp. 63-64.
5   Möllhausen, *Reisen*, chapter 13.

23D-E
Mohave Canyon in March 1990

CAT. NO. 24

Balduin Möllhausen
The Needles, View from the North
Watercolor and gouache on paper
7¾ x 11 in. (19.7 x 27.9 cm)
signed l.r.: "Möllhausen"
l.l. on paper mount: "47"
l.c. on paper mount: "The Needles view from the north"
ACM 1988.1.45

Cat. No. 24

Möllhausen sketched another view of the Needles from the north, near the present town of Topock in Mohave County, Arizona (cat. 24b).[1] He had visited this area once before, when Whipple's expedition camped nearby on the nights of February 24-25, 1854, after detouring to the east around the Needles. A railroad bridge stands today at the location Whipple recommended, "where gentle slopes of the prairie form low bluffs impinging upon the river."[2] The Ives expedition reached this vicinity on the morning of February 10, 1858, after leaving their camp in Mohave Canyon. Möllhausen wrote: "We were still in the canyon after travelling two miles, but the rocks were not as large. The river became wider, the mountains receded completely, and the broad valley of the Mohave Indians opened up before us." Later, he noted, "when we looked behind us, we could see the entire Needles chain with its pointed towers and sharp spires."[3]

The lithograph in Ives' *Report* (cat. 24a) retains the general forms and outlines of the Needles, the scrub vegetation along the riverbank, and the bathing Mohave natives in the foreground, all found in Möllhausen's sketch. The lithographer for the most part even achieved the same tonal contrasts found in the watercolor. However, he omitted the driftwood log and snags, and by reducing the size of the Indians, effectively increased the relative size of the distant peaks. *Reisen* also contains a print after this sketch; it basically adheres to the original but substitutes a pelican and a herd of pronghorn antelopes for the bathing Mohaves.[4]

24A
John James Young after Balduin Möllhausen, *The Needles (Mojave Range)*,
toned lithograph, from Ives, *Report* (1861), part 3, opp. p. 30,
Amon Carter Museum Library

24B
The Needles from the north, viewed from near the railroad bridge
in Topock, Arizona, March 1990

The presence of Indians in Möllhausen's watercolor is consistent with the experiences the travellers reported at that time. As noted in cat. no. 12, Lieutenant Ives and his men saw Mohaves in large numbers along the river (and diving for beads) on January 22, 1858. On January 29, before they entered the Monument and Mohave canyons, Mollhausen reported: "large Indian settlements appeared to lie hidden behind the willow woodlands. Natives of both sexes and all ages happily bustled about on the sandy bank which extended in terraces far out into the stream, or lay sunning themselves on the soft ground." The following day, a "crowd of Mohave Indians were assembled on the bank early in the morning,"[5] and on February 1, Lieutenant Ives and Captain Robinson came upon two men whom Ives recognized at once as Mohaves:

One of them must have been nearly six feet and a half in height, and his proportions were herculean. He was entirely naked, excepting the ordinary piece of cotton about his loins, and his chest and limbs were enormously developed. A more scowling, sinister looking face than that which surmounted this noble frame I have seldom seen; and I quite agreed with a remark of the captain, that he would have been an unpleasant customer to encounter alone and unarmed. His companion was smaller, though a large man, and had a pleasant face.[6]

As the steamer emerged from Mohave Canyon on February 10, the party saw more Mohaves beginning to cluster on the banks. Ives wrote that he

was glad to see, from the presence of the women and children, that they had no immediate hostile intentions. A chief, with a train of followers in single file, approached the edge of the bank to pay his respects, but as it was not convenient just then to stop, I made signs to him to visit us in camp at evening. All day the Indians have followed us, examining the boat and its occupants with eager curiosity. They, on their side, have been subjected to critical inspection, which they can stand better than any of the tribes that live below.[7]

Ives and his expedition had better opportunities to observe the Mohaves when large crowds of both men and women visited them in camp on February 11 (camp 42), eager to trade beans, corn, wheat, and pumpkins.[8] Although Möllhausen noted strong similarities between the various Colorado River tribes,[9] Ives commented

24c
John James Young? after Balduin Möllhausen, *Mojaves,* toned lithograph
(black with red and gold), from Ives, *Report* (1861), part 1, opp. p. 66,
Amon Carter Museum Library

24d
After Balduin Möllhausen, *Mojave Indians,* toned lithograph
(black with gray and brown), from *U.S.P.R.R. Explorations and Surveys,*
vol. 3, part 3, opp. p. 33

on the gigantic stature of the Mohave men, most of whom had "noble figures, . . . intelligent countenances and an agreeable expression." Although the younger girls "are very pretty and have slender, graceful figures," women "over the age of eighteen or twenty, are almost invariably short and stout, with fat, good-natured faces. Their only article of dress is a short petticoat, made of strips of bark, and sticking out about eight inches behind. . . The children wear only the apparel in which they were born, and have a precocious, impish look." Babies "have a sharp, wide-awake expression, and their faces may always be seen peering from under their mother's arms, spying out what is going on. . . . It is rare for one of them to utter a cry, which may be attributed to the judicious system of their early training."[10] Ives also noted that minor chiefs wore a crimson-tipped white plume and inferred that rank must be somewhat hereditary, since he had seen a boy wearing such a "badge of distinction."[11]

Möllhausen's numerous renderings and observations of Mohaves form some of the most important ethnological material that he gathered while in America. He was not the first artist to depict them: Richard H. Kern, the artist of the Sitgreaves expedition, had sketched the Mohave as early as 1851 (see fig. 33).[12] Möllhausen first encountered members of this tribe in late February 1854 with Whipple's expedition, and recorded many observations on their customs, including several sketches of a Mohave dwelling or earth lodge and of food storage containers.

During the 1858 trip upriver, the explorers on the steamboat continually had friendly encounters with the Mohaves. Chief Cairook and Iretéba, two well-known Mohaves who had guided the Whipple expedition, again provided their valuable services to Ives, and both Möllhausen and the lieutenant wrote in high praise of their friendship. Iretéba and several Mohave guides remained with Ives on the overland portion of the expedition until April 5.[13] In *Mojaves,* a lithograph after Möllhausen in Ives' *Report,* Iretéba appears at left and Cairook in the center (cat. 24c). At right stands a woman with a child on her hip. The three-figure format and subject matter are quite similar to several of Möllhausen's earlier works from the Whipple expedition, especially the lithograph titled *Mojave Indians* from *Explorations and Surveys* (cat. 24d). In both lithographs two tall Mohave men, wearing only breechclouts, stand at left and center, with the man at left holding a bow and arrows.

Both prints also depict a short but well-proportioned woman at right, wearing only a bark skirt and holding a reed basket.

In addition to these prints, Möllhausen made several watercolors of Mohaves and their customs; two artworks are now in the Berlin Museum of Ethnology. An excellent finished watercolor by Möllhausen, titled *Mohave Indianer, Col. Exp. 1857-58* (see plate, p. 131), shows four adult figures: a warrior shooting an arrow with his bow, a woman with a child on her hip and a basket of food on her head, a seated man with an elaborate headdress, and a warrior leaning on a pole or spear. The natives strike convincing poses and form a somewhat less formal group than those in the lithographs. Möllhausen originally may have intended for this watercolor to illustrate *Reisen* but decided against it because of cost.

A second watercolor shows a favorite Mohave pastime (see plate, p. 131). In 1854 Mollhausen noted that almost all the men carried long poles and were untiring in playing a ring game. In 1858, he also observed:

I learned here for the first time that the game, in which two men run at full speed and throw poles at a rolling ring, is not just for entertainment and a pastime, but is also associated with gambling. Our interpreter, Maruatscha, who always had a small supply of our beads and sundry glittering things at his disposal for services he had performed, wagered them whenever he could compete in the game on the bank with his countrymen. Each time he returned to the boat without them, proof that the Mohave Indians surpassed him in dexterity. . . . [O]ne day [his] red shirt had disappeared, and Maruatscha returned . . . stark naked. He had gambled away his entire possessions except for his loin cloth and a nose ring.[14]

The details of the watercolors and lithographs of the Mohaves largely agree with contemporary eyewitness descriptions and surviving artifacts from the period. Europeans and white Americans generally favored the tribe's physical appearance over that of other Colorado peoples. Ives described Cairook as "nearly six feet and a half high," with "a magnificent figure and a fine open face."[15] Möllhausen commented on Cairook's "powerful robust physique," adding that "his movements were extremely graceful and well coordinated. He smiled constantly, expressing his innately happy disposition."[16] A soldier who saw Iretéba (cat. 24e) around 1863

24E
Artist unknown,
Irataba, Chief of the Mojaves,
engraving, from *Harper's Weekly,*
February 13, 1864

wrote similarly: "Iraytba [*sic*] was a man six feet four inches tall, of very powerful frame, but very gentle and kind in demeanor, and a staunch friend of the whites," while a writer for *Harper's Weekly* called him "the finest specimen of unadulterated aboriginal on this continent."[17] Möllhausen noted that when he saw Iretéba in 1858, he had cut his hair shorter than normal for Mohave men and women because he was in mourning for a brother who had been killed in a raid against the Cocomaricopa tribe.[18]

One of Cairook's wives rode along with her husband and Iretéba on the *Explorer* on February 15, and she could have served as the model for the woman in the lithograph from Ives' *Report* (cat. 24c). Möllhausen noted:

She was about twenty-eight years old, and she had more captivating features than I had ever seen among the natives. She did not look like an Indian. One could have taken her for a European in brown make-up, if the rest of her appearance had not been taken into consideration. Her dress and appearance were like those of other Indian women; her thick black

24F

After Balduin Möllhausen, *Ceremonial Visit of Me-sik-eh-ho-ta,*
toned lithograph (proof), Whipple Collection, Oklahoma
Historical Society, photo 9256-UL5

hair reached from her forehead only to her eyebrows, while in
the back it reached her shoulders. Her lower lip and her lower
eyelids were tattooed a dark blue, and between the corners of
her mouth her chin was decorated with blue stripes and dots.
Her well-developed upper body was not clothed or tattooed. A
skirt fashioned from thick strips of bark fell from her hips to
her knees, and bestowed upon this woman a graceful, even
striking appearance.[19]

Despite the similarity of this description to the woman in the
lithograph, however, there are no tattoos on the woman in the
picture, and neither Möllhausen nor Ives mentions Cairook's wife
having a child with her.[20]

Both body paint and facial tattooing distinguish the natives in
the earlier Whipple lithograph and *Mohave Indianer.* Möllhausen
described four unidentified native men he encountered along the
Colorado with Whipple as having faces "painted, in a really terrific
manner, coal-black, with a red streak passing from the forehead
over nose, mouth, and chin: a style of decoration that must be very

fashionable among these Indians, as I afterwards saw it frequently."
The women, he noted, "go more carefully to work with their
painting than the men, and tattoo themselves more; their lips are
mostly coloured quite blue, and their chin, from one corner of the
mouth to the other, is adorned with blue lines and spots."[21] The
man at left in the Whipple lithograph (cat. 24d) may have paint in
his hair; according to Möllhausen, the men sometimes adorned
themselves with white, blue, red, or yellow.[22]

One of the most curious details in Möllhausen's watercolor of
the four Mohaves (see p. 131) is the large headdress worn by the
seated man. To a casual observer it looks like something belonging
only on a Plains Indian, but the headdress may indeed be Mohave
and similar to one Möllhausen saw in 1854. As Whipple's men
were preparing to cross the Colorado River, they met "an aged
chief, Me-sik-eh-ho-ta, a venerable-looking man, with an immense
plume of feathers on his head, and a thick spear in his hand."[23]
Me-sik-eh-ho-ta, with his unusual headdress, appears in a rare
lithographic proof (cat. 24f) after a lost 1854 sketch by Möllhausen;
the headdress is very similar to the one in the Berlin watercolor,
dating from soon after Ives' expedition, suggesting that both
images were drawn from direct observation.[24]

In August 1858, shortly after the Ives expedition visited the
Mohaves, hostilities erupted between the tribe and a wagon train of
white settlers along Beale's wagon road. This led to the permanent
introduction of U.S. troops into Mohave territory and the impris-
onment in 1859 of several Mohave chiefs, among them Cairook, at
Fort Yuma. About two months later, a guard bayoneted Cairook as
he and several others attempted an escape. After Cairook's death,
the whites recognized Iretéba as the principal chief of the Mohaves.
He visited Los Angeles in 1860-61 and again in 1863, after gold
was discovered in Mohave territory and a flux of prospectors had
moved in. On the later trip, he travelled by steamer to San Francisco, New York, Philadelphia, and Washington, D.C., before
returning to his people. Iretéba died at his home near La Paz,
Arizona, in 1874.[25]

---

1   See USGS 1:250,000 map "Needles."
2   *Explorations and Surveys,* vol. 3, part 2, p. 41. See also part 1, p. 113, and
    Foreman, *A Pathfinder in the Southwest,* pp. 234-235. Whipple's men passed
    around the Needles on February 24 and then returned to the river's edge.

On both days they were in contact with the Mohaves. Albert H. Campbell, Engineer and Surveyor for Lieutenant Whipple's expedition, wrote in his preliminary report in *Explorations and Surveys*, vol. 3, intro., p. 26: "Following up the left bank of the Great Colorado, whose ascent for thirty-four miles is about one and a half foot per mile, a suitable point for crossing was found among the "Needles," a series of porphyritic and trap dikes, through which the stream forces a passage. Notwithstanding the formidable appearance of the rocks at a casual glance, there are but three points where they infringe directly upon the river, and these points are quite narrow and easily perforated or blasted off entirely." In addition to the railroad, Interstate 40 (old U.S. Highway 66) also crosses the river here. According to Granger, *Arizona's Names*, p. 621, the name "Topock" is a Mohave word of unknown meaning applied to a site on the California side.

3   Möllhausen, *Reisen*, chapter 13. Also see Ives, *Report*, part 1, pp. 60, 63, and Egloffstein's map 1. According to Dr. Newberry, the Needles themselves were "formed of purple porphyry and trachyte"; Ives, *Report*, part 3, p. 30.

4   See *Die Nadelfelsen oder Needles (von Norden gesehen)*, in *Reisen*, vol. 1, between pp. 274 and 275.

5   Möllhausen, *Reisen*, chapter 11.

6   Ives, *Report*, part 1, p. 59.

7   Ives, *Report*, part 1, p. 66.

8   Ives, *Report*, part 1, pp. 66-67.

9   Many descriptions of the Cocopa, Yuma, and Chemehuevi by Möllhausen and his contemporaries are almost interchangeable with their descriptions of the Mohave, the northernmost and largest of the Yuman-speaking tribes of the lower Colorado River. See Möllhausen, *Diary of a Journey*, vol. 2, pp. 244-247, and Kenneth M. Stewart, "Mohave," in Sturtevant and Ortiz, eds., *Handbook of North American Indians*, vol. 10, *Southwest*, pp. 55-70.

10 Ives, *Report*, part 1, p. 66. Ives further noted of Mohave infant care: "Those that are very young the mothers, with unusual good judgement, dispose of by tying them in a wooden arrangement, shaped like an old fashioned watch case, which may be carried in hand and at the same time conveniently put away till required for nursing. When a few months older, they are taken out of the case and carried upon the projecting petticoat, where they sit astraddle, with their legs clasping their mother's waist and their little fists tightly clutched in her fat sides. They nurse without moving their position, having only to elevate their mouths at a slight angle."

11 Ives, Report, p. 67.

12 Weber, *Richard H. Kern*, pp. 177-178, 321 n.59. The principal figures in Kern's image were two tall Mohave men and a woman, more imaginatively posed than those in the lithographs after Möllhausen.

13 Ives, *Report*, part 1, pp. 69-70; see also Whipple, "Report," in *Explorations and Surveys*, vol. 3, part 1, pp. 119-121.

14 *Reisen*, chapter 11. In his earlier *Diary of a Journey*, vol. 2, p. 259, Mollhausen noted something of this ring game as he witnessed it in 1854:

Several of the men carried poles sixteen feet long in their hands, the use of which I could not make out, till I saw the brown forms leave the crowd two by two, to begin a game, which remained somewhat obscure to me, though I looked on at it for a long time. The two players placed themselves near one another, holding the poles high up, and one of them having in his hand a ring, made of strips of bast, of about four inches in diameter. Lowering the poles, both rushed forward, and at the same time the one who held the ring rolled it on before him, and both threw the poles, so that one fell right and another left of it, and arrested its course. Without stopping a moment, they then snatched up the ring, and the poles, and repeated the same movements back again, over the same spot, a piece of ground about forty feet long, and so on again and again, till the indefatigable players had trampled a firm path on the loose soil of the meadow. They continued this game for hours without stopping a minute, or exchanging a single word, and though some of the Indian spectators joined them, they were just as much absorbed in the game, as the players themselves, and would by no means allow me to come nearer, to try and make out the meaning of it. . . . it was evident here and elsewhere, that these Indians were as passionately in earnest about this game, as the most enthusiastic chess player could be amongst us.

The Oklahoma Historical Society has an earlier Möllhausen watercolor of this subject, titled *Two Indians Playing Pole and Ring Game* (*Game — Two Indians in figure*), in its Whipple Collection; another version of the Berlin watercolor is in the Neues Palais Collection, Potsdam.

15 Ives, *Report*, part 1, p. 69.

16 Möllhausen, Reisen, chapter 14. A lithograph titled *Indian Ornaments and Manufactures* in Whipple's Indian report has a picture of a nose ornament (figure E) worn by Cairook, consisting "of a large white bead of shell, and from it hangs a thin conical slip of bright and light-blue stone. The small leather thongs were passed through the septum of the nose, and secured the gem to it." Whipple et al., "Report Upon the Indian Tribes," in *Explorations and Surveys*, vol. 3, part 3, p. 51; lithograph between pp. 50 and 51. Möllhausen recorded in 1858 that Cairook had "a thong through the cartilage of his nose on which a large white pearl and a blue turquoise stone were fastened"; *Reisen*, chapter 14. Cairook evidently wears such an ornament in the lithograph for Ives' *Report*. Necklaces worn by all three figures in the print and the woman in the watercolor may be either of native or trade origin.

Other items depicted in the *Indian Designs & Manufactures* lithograph also resemble articles in the Ives lithographs and watercolor: a two-feathered headdress of Mohave origin (figure 9), a Yampai bow (figure 1), a Mohave lance (figure 7), a multi-colored Mohave apron (figure 12) and bark aprons and cord (figures 10, 11, and 13), and a basket (figure 14) described as of Paiute origin. Möllhausen repeated the image of a Mohave woman with a basket on her head several times. Engraved plate 25 after Möllhausen in Whipple's "Report upon the Indians" shows a Mohave woman in a bark skirt with a similar basket on her head and a Mohave man in a breechclout with a multi-feathered headdress holding a spear—and may have been inspired by the lithograph of Mohave Indians after Richard Kern in Sitgreaves' report (see fig. 33).

17 Edward Carlson, "The Martial Experiences of the California Volunteers," *Overland Monthly* (May 1886), pp. 480-486, and *Harper's Weekly*, Feb. 13, 1864, both quoted in Arthur Woodward, "Irataba—Chief of the Mohave," *Plateau, Museum of Northern Arizona* 25, no. 3 (January 1953): 53.

18 Möllhausen, *Reisen*, chapter 14.

19 Möllhausen, *Reisen*, chapter 15. See also Ives, *Report*, part 1, p. 70.
20 The woman and clinging child in the Berlin watercolor and the Ives *Report* lithograph are similar (perhaps coincidentally) to a Möllhausen field sketch of a Native American mother with a small child on her shoulder and another young child clinging to her. The woman in the sketch has her hair cut like a lower Colorado River woman and wears a bark apron or skirt with a wide cord or belt similar to the one in the *Mojaves* lithograph; she also wears a blanket or cloak of some kind over her shoulders. Möllhausen combined such field sketches to create his finished, multi-figural compositions.
21 Möllhausen, *Diary of a Journey*, vol. 2, pp. 242, 245. Early photographs illustrating Mohave women with tattooing appear in Stewart, "Mohave," in Sturtevant and Ortiz, eds., *Handbook of North American Indians, Volume 10, Southwest*, p. 63.
22 Möllhausen, *Diary of a Journey*, vol. 2, p. 250.
23 Möllhausen, *Diary of a Journey*, vol. 2, pp. 265-266. There is an outstanding seated portrait of Me-sik-eh-ho-ta in the Neues Palais Collection, Potsdam.
24 Whipple wrote of Mohave men wearing eagle feathers, and Möllhausen thought that at least some also wore swan, vulture, or woodpecker feathers. The physiognomies, hairstyles, headdresses, body tattoos, and facial paint in these images look reasonably accurate. Some of the men carry long poles used in the ring and pole game. Möllhausen also wrote about an occasional horse among the Mohaves, as seen at left in cat. 24f, and of demonstrations of affection between all, irregardless of sex, as seen by the two men embracing, just to the left behind Me-sik-eh-ho-ta. Only the strange hat on the man at far left and the tunic on the boy or girl at far right seem out of place.
25 Information on Iretéba and Cairook's later activities comes from Woodward's fascinating article, "Irataba," pp. 53-68. Woodward constructed the biography from many sources, including contemporary newspaper accounts and the reminiscences of Iretéba's granddaughter, but he was apparently unaware of the fine accounts of the two Mohaves in Möllhausen's *Reisen*.

Cat. No. 25

Balduin Möllhausen
Boundary Hill (Boundary Cone)
Watercolor and gouache on paper
8⅛ x 10⅝ in. (20.6 x 27.0 cm)
signed l.r.: "Möllhausen"
l.l. on paper mount: "34"
l.c. on paper mount: "Boundary hill"
ACM 1988.1.32

Cat. No. 25

The southern tip of the present state of Nevada reaches the Colorado River very near the point where Möllhausen made the sketch for his watercolor titled *Boundary Hill*. However, instead of facing northwest toward Nevada or west toward California, the view faces east, toward the Arizona side of the river. The formation, usually known as Boundary Cone, is a prominent landmark in the region.[1]

The expedition reached this point on February 16, just after leaving camp 45. Möllhausen noted this "eastern gravel plain" and wrote that they

> could see the ascending desert to the east all the way to where it stopped at the edge of a cluster of mountains. These mountains [Black Mountains] were joined together and appeared to be a long mesa. The plain itself had a gray surface, the result of sterile soil which lacked the delicate green cover of spring and the blossoming flowers of summer. It extended far away like a picture of death, its color gradually shading until it blended with the blue mountains.[2]

Möllhausen described a thin strip of vegetation along the riverbank precisely as he painted it in his watercolor:

> Various varieties of willows grew in pleasant groves on the narrow ribbon of sand between the river and the desert. Isolated cottonwood trees rose high above them with fantastically formed tops resting on naked trunks. From a distance

they almost looked like ornamental trees which had been artificially grown and pruned.[3]

These "shimmering dense green forms . . . contrasted sharply with the arid wasteland," as he managed to convey pictorially.

Möllhausen may have taken his view with the aid of a telescope. In reality, the Black Mountains and Boundary Cone appear much farther away than he depicted them, but the basic outlines of the formations are accurate (see cat. 25b). The two Indians in the right foreground may represent actual Mohaves, but their principal function is to add a sense of scale to the picture.

Although Ives' *Report* contains an engraving after Möllhausen's watercolor (cat. 25a), the lieutenant made no mention of the formation. The engraver embellished and distorted the shapes of the mountains in the distance and altered the direction of the falling light.[4]

Möllhausen may have inadvertently included Boundary Cone in one of his sketches for the Whipple Expedition, which crossed

25B
Boundary Cone in 1990

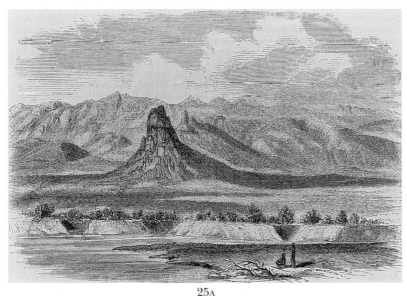

25A
John James Young after Balduin Möllhausen,
Boundary Cone, engraving, from Ives, *Report* (1861), part 1, p. 71,
Amon Carter Museum Library

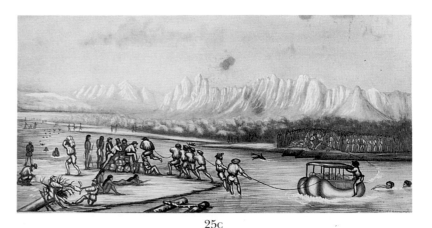

25C
Balduin Möllhausen, *Whipple's Expedition Crossing the Colorado River (Rio Colorado near the Mohave Villages)*, watercolor and opaque white with graphite on brown paper mounted on buff paper, Whipple Collection, Oklahoma Historical Society, photo 9256-P70

the Colorado River from east to west on February 27, 1854 (cat. 25c). Using Mohave reed rafts and an old, worn India rubber and canvas pontoon boat to aid them in their crossing, Whipple reported:

> our loaded barge [capsized] nearly three times casting its contents into the river. Mr. White and a little Mexican boy were nearly drowned, before the exertions of Mr. Möllhausen succeeded in extricating them from beneath the boat. The Indians, who are capital swimmers, plunged in saving much of the property.[5]

Möllhausen recorded the event from an island in the middle of the river; Boundary Cone may be distinguished at far right in the picture.[6] Although neither Whipple's "Report" nor Möllhausen's *Diary of a Journey* specifically mention the cone, it is clear that the crossing took place about thirteen miles from it, and this prominent landmark would have been visible.

---

1 Compare USGS 1:250,000 map "Needles" and Egloffstein's map 1; also see Granger, *Arizona's Names*, p. 86.

2 Möllhausen, *Reisen*, chapter 15.

3 Ibid.

4 Boundary Cone also appears in the center of the panoramic view by Baron Egloffstein titled *Colorado River, Mojave Valley. From Mojave Mountains to Pyramid Cañon*; see Ives, *Report*, part 1, opp. p. 64. To obtain this view, Egloffstein apparently climbed the foothills of what are today known as the Dead Mountains in San Bernardino County, California. The panorama gives a fairly accurate impression of the terrain; compare it with USGS 1:250,000 map "Needles."

5 Whipple, "Report," in *Explorations and Surveys*, vol. 3, part 1, pp. 116-118. See also Möllhausen, *Diary of a Journey*, vol. 2, pp. 265-272. Lieutenant Ives had brought the pontoon boat with him from Texas.

6 The lithograph after Möllhausen for Whipple's topographical report, titled *Rio Colorado Near the Mojave Villages. View No. 2. from an Island Looking North.*, does not show Boundary Cone as clearly, and neither did Möllhausen's watercolor of the same subject, which was formerly in the Berlin Museum of Ethnology. Two other lithographed views after sketches by Albert H. Campbell depict the crossing, but were taken from angles which would not show Boundary Cone. They are *Rio Colorado — Near the Mojave Villages. View No. 1. from the Left Bank Looking W. N. W.* and *Rio Colorado Near the Mojave Villages. View No. III. from the Right Bank, Looking East.* All three lithograph versions were redrawn by John James Young for the frontispieces of *Explorations and Surveys*, vol. 3, parts 2, 1, and 3, respectively.

CAT. NO. 26

Balduin Möllhausen
Beale's Crossing (Beale's Pass)
Watercolor and gouache on paper
7¾ x 11¼ in. (19.7 x 28.6 cm)
signed l.r.: "Möllhausen"
l.l. on paper mount: "25"
l.r. on paper mount: "Beale's Crossing"
ACM 1988.1.24

Cat. No. 26

Möllhausen's important watercolor titled *Beale's Crossing* or *Beale's Pass* depicts an actual location with topographical accuracy, but portrays an event that the artist did not witness himself. This view, from a sketch made near present-day Riviera, Arizona, faces east northeast toward Cathedral Rock and Battleship Mountain in the Black Mountains; today the ruins of Fort Mohave (established in 1859) stand in this vicinity, on gravel bluffs like those on the left bank in the picture.[1] On February 16, 1858, Möllhausen wrote, Ives' expedition "landed on the left bank and set up camp under beautiful cottonwood trees which grew close together in a forest." Such trees, identifiable by their bare trunks, line the riverbanks in this watercolor. Möllhausen noted that this camp (no. 46) was "only a short distance from the point where Lieutenant Beale and his camel expedition had crossed the Colorado."[2]

U.S. Navy Lieutenant Edward Fitzgerald Beale (1822-1893) (cat. 26c) actually crossed the Colorado at this point on at least two occasions. While he was constructing a wagon road to California, Beale arrived here from the east on October 18, 1857, with a herd of camels that he was testing for the War Department. The next day his men transported their wagons and baggage across the river in an India-rubber boat, and the camels and mules followed a day later. Beale was concerned that the camels would not swim, and in fact,

[t]he first camel brought down to the river's edge refused to take to the water. Anxious, but not discouraged, I ordered another one to be brought, one of our largest and finest; and only those who have felt so much anxiety for the success of an experiment can imagine my relief on seeing it take to the water, and swim boldly across the rapidly flowing river. We then tied them, each one to the saddle of another, and without the slightest difficulty, in a short time swam them all to the opposite side in gangs, five in a gang; to my delight, they not only swam with ease, but with apparently more strength than horses or mules.[3]

Möllhausen learned about Beale's camels while in Los Angeles, preparing for Ives' expedition. On November 9, 1857, he wrote:

a number of suntanned figures rode in on dromedaries from the direction of San Bernardino. Their appearances exhibited the unmistakable signs of a long journey through desolate regions. It was Lieutenant Beale, formerly of the United States

26A
John James Young after Balduin Möllhausen, *Beale's Pass*, engraving, from Ives, *Report* (1861), part 1, p. 74, Amon Carter Museum Library

Marines, who, under orders from his government, had made the first attempt to journey across the American prairies and mountains on the African "ships of the desert."

We were able to exchange only brief greetings, because preparations for our departure required so much time and Lieutenant Beale was so busy discharging his people. . . Since our mutual goal was Fort Tejon where the expedition's twenty-two camels were to be wintered. . . , I had several opportunities to learn more about the introduction of camels in America.[4]

While in Los Angeles or at Fort Tejon, California, Möllhausen may have made a portrait of himself sitting on a surprised mule, with a camel and a faint adobe building in the background (cat. 26b). (He might also have made this sketch sometime after the Ives expedition, using other images or studies made on a visit to the Berlin Zoo.)

On January 1, 1858, Lieutenant Beale and his men left Fort Tejon with camels, wagons, and mules, heading back east. As Beale approached his Colorado crossing on January 23—just weeks prior to Ives' arrival here—he learned to his great surprise from an advance scout that the steamer *General Jesup*, with its owner Captain Johnson at the helm and a military escort under Lieutenant James L. White, was by coincidence already at the crossing and could convey the expedition to the opposite side. Beale wrote in his journal:

It is difficult to conceive the varied emotions with which this news was received. Here in a wild almost unknown country, inhabited by savages, the great river of the West hitherto declared unnavigable had for the first time borne upon its bosom that emblem of civilization, a steamer. . . . But alas! for the poor Indians living on its banks and rich meadow lands. . . [t]he steam whistle of the *General Jesup* sounded the death knell. . . .

In a few minutes after our arrival the steamer came alongside the bank and our party was transported at once with all our baggage to the other side. We then swam the mules over and bidding Captain Johnson good-bye he was soon steaming down the river towards Fort Yuma three hundred and fifty miles below.[5]

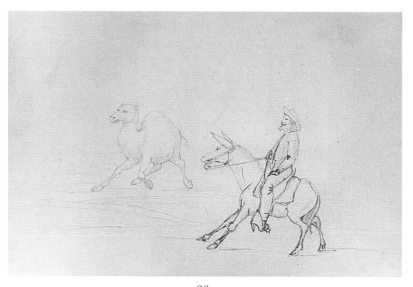

26B
Balduin Möllhausen, *Möllhausen's Mule Alarmed at the Sight of a Camel
[Möllhausen zu Pferd, daneben ein Kamel]*, graphite on paper, from
Sketchbook 3, Möllhausen Family, Bleicherode, Germany, photograph
courtesy of Dr. Friedrich Schegk, Munich

26C
This portrait of Edward Beale hangs today in Decatur House,
Washington, D.C.

Beale added that the presence of his camels, "hundreds of wild
unclad savages," the dragoons of his escort, and the steamer, had
made "a curious and interesting picture."[6]

The crossings evidently left an impression on the natives as
well. Möllhausen related that a Mohave called Captain Jack, "who
especially liked to travel on board the steamboat and who therefore
tagged along with us, described with considerable detail how the
'large long-necked horses' crossed the river, for he had been there
from beginning to end."[7]

Regarding his meeting with Captain Johnson and the *General
Jesup*, Lieutenant Beale wrote: "I confess I felt jealous of his achieve-
ment and it is hoped the government will substantially reward the
enterprising spirit which prompted a citizen at his own risk and at
great hazard to undertake so perilous and uncertain an expedition."[8] Unfortunately, Lieutenant Ives shared Beale's jealousy of
Johnson without the accompanying magnanimity. Ives omitted
almost all references to Johnson in his *Report Upon the Colorado River*

of the West in order to emphasize the exploratory nature of his own achievements, and he apparently tried to prevent Johnson from receiving the government recognition that was due him. Arthur Woodward, in his book *Feud on the Colorado*, speculated that Ives, through his wife's connection as niece to Secretary of War Floyd, may have had Lieutenant James L. White's report of Johnson's voyage pigeon-holed in order to prevent it from reaching publication.[9]

Ives briefly mentioned Beale in his own description of the area, written on February 17:

> The range east of the Mojave valley we call the Black mountains. These mountains run from a point fifteen or twenty miles east of the foot of the valley in a northwesterly direction, and cross the Colorado about fifty miles north of camp. . . . Westward, opposite to camp, is the pass through the spur that connects the Black and Mojave ranges, by which the wagon trail of Lieutenant Beale leaves the valley of the Colorado.[10]

However, he may have ordered the engraver of his *Report* to omit Möllhausen's camels from the print—again with the intent of maximizing the exploratory nature of his own expedition. Instead of showing Beale's expedition, the engraver compressed the shape of the mountain in the center right and substituted a raft with two Indians and a dog (cat. 26a).

Ives and his men camped near Beale's Crossing again on the nights of March 21 and 22, on their return from the upper Colorado and after their rendezvous with the mule train from Fort Yuma. From here on March 23, Lieutenant Ives sent the *Explorer* back downriver while he and many of the men continued their explorations overland, to the northeast.[11]

---

1   Compare Egloffstein's printed map; original map, plat no. 40; USGS 1:250,000 map "Kingman"; USGS 15' series map "Davis Dam Quadrangle"; and USGS 7.5' series "Davis Dam SE" and "Oatman" quadrangles. For information on Fort Mojave, see Altshuler, *Starting with Defiance*, pp. 43-47.

2   Möllhausen, *Reisen*, chapter 15.

3   Beale to John B. Floyd, Secretary of War, letter begun and dated October 18, 1857, quoted in Stephen Bonsal, *Edward Fitzgerald Beale, A Pioneer in the Path of*

Empire, 1822-1903 (New York and London: G.P. Putnam's Sons, 1912), p. 216. Compare also the October 19 and 20 entries from the journal of May Humphreys Stacey, published in Lesley, *Uncle Sam's Camels, The Journal of May Humphreys Stacy, Supplemented by the Report of Edward Fitzgerald Beale (1857-1858)* (Cambridge: Harvard University Press, 1929), pp. 114-115. Stacey recorded that the camels all got across safely, but that ten mules and two horses drowned.

    Elliott Coues, historian and former army surgeon at Fort Mohave, speculated that Spanish priest Padre Francisco Garcés crossed the Colorado somewhere near this location on June 4, 1776, heading east-northeast. Coues also believed that Garcés crossed the Black Mountains (which the latter called the Sierra de Santiago) by the same pass that Beale and Ives used. See Coues, ed., *On the Trail of . . . Garcés*, pp. 312, 314.

4   Möllhausen, *Reisen*, chapter 3.

5   Entry for January 23, 1858, in "The Report of Edward Fitzgerald Beale to the Secretary of War Concerning the Wagon Road from Fort Defiance to the Colorado River, April 26, 1858," 35th Congress, 1st Session, House of Representatives, Ex. Doc. 124, reprinted in Lesley, *Uncle Sam's Camels*, part 4, pp. 261-262. Bonsal, *Edward Fitzgerald Beale*, pp. 227-228, quotes extracts from Beale's report.

6   Beale's report, January 23, 1858, in Lesley, *Uncle Sam's Camels*, p. 262, and in Bonsal, *Edward Fitzgerald Beale*, p. 228.

7   Möllhausen, *Reisen*, chapter 15.

8   Beale's report, January 23, 1858, in Lesley, *Uncle Sam's Camels*, p. 262, and in Bonsal, *Edward Fitzgerald Beale*, p. 228.

9   Woodward, *Feud on the Colorado*, pp. 69-130.

10 Ives, *Report*, part 1, p. 74. Ives mistakenly believed Sitgreaves had reached the Colorado from the east through this pass, but actually Sitgreaves had crossed the Black Mountains at Union Pass. Sitgreaves' men descended the east bank of the river from present-day Bullhead City, and made their camp 33 in the river bottom below the hill on which Fort Mojave later stood. See Wallace, "The Sitgreaves Expedition," pp. 353-354, and Granger, *Arizona's Names*, pp. 568, 644.

11 Möllhausen, *Reisen*, chapters 15 and 19.

CAT. NO. 27

Balduin Möllhausen
Obelisk Mountain, View Down the River (Pyramid Cañon)
Watercolor and gouache on paper
8⅜ x 10⅝ in. (21.3 x 27.0 cm)
signed l.r.: "Möllhausen"
l.l. on paper mount: "36."
l.c. on paper mount: "Obelisk m./view down the river."
ACM 1988.1.34

Cat. No. 27

From the camp near Beale's Crossing, the *Explorer* continued its way up the Mohave Valley on February 17 and 18, reaching five miles north of present-day Bullhead City, Arizona, on the second morning. There, Lieutenant Ives wrote, "the river narrowed and entered a cañon through a gate, one side of which looked like the head of a bull."[1]

Davis Dam now stands just below this gate, formed where the river, flowing from the north, cuts through the junction of the Newberry Mountains (named in honor of the geologist) on the west bank and the Black Mountains on the east bank.[2] George A. Johnson and Lieutenant James L. White, who had reached this canyon in mid-January—a few weeks earlier than Ives—in the steamboat *General Jesup*, named it El Dorado Canyon.[3]

Ives, Newberry, and Möllhausen all wrote that the scenery in this canyon, although beautiful and of considerable interest, could not compare with that of several other canyons they had already seen. They called the passage Pyramid Canyon for "a remarkable monument-like pinnacle of porphyritic rock which crowns the left bank near the northern entrance."[4] Lieutenant Ives described the formation as "a natural pyramid, of symmetrical proportions, twenty or thirty feet high"; Egloffstein labelled it on his original maps as "loaf" and "Loaf rock," and this name has remained.[5] It appears on the left in Möllhausen's watercolor view, taken from a sketch he made looking downriver. The artist preferred to call the pyramid an "obelisk" or "Tower Rock," a name his watercolor and description tend to support:

Now and then the volcanic rocks formed steep banks on both sides and exhibited formations resembling pillars. We conferred the name Obelisk Mountain on one of these formations, because of its uniform shape. In fact, the rock looked very much like a monument when seen from the south, but had I first seen this point from the north, I believe the name Tower Rock would have been more descriptive. It only required a little imagination to see the round pillar as a watch-tower with a half-open gate.[6]

The artist added that rapids near this formation and throughout the canyon made it "one of the most dangerous places on the river that we encountered during the entire journey." Yet Captain Robinson skillfully guided the *Explorer* through.

The expedition passed through Pyramid Canyon again while descending the river the following month. Ives' entry for March 20 notes that they camped "at the foot of Pyramid cañon, under the lee of the Bull's Head rock," where the pack train of mules com-

27A
John James Young after Balduin Möllhausen, *Pyramid Cañon*, engraving, from Ives, *Report* (1861), part 1, p. 76, Amon Carter Museum Library

manded by Lieutenant Peacock had already arrived.[7] Since the watercolor by Möllhausen faces downriver, it is possible that he made the original sketch for this view on the return trip.

The unusual and anthropomorphic names that the nineteenth-century explorers gave to topographical landmarks along the Colorado River may indicate the romantic perceptions with which they viewed the landscape. In this particular watercolor, Möllhausen stretches his credibility as a faithful delineator of the topography by including a weird, unearthly formation at right. Interestingly, the engraver corrected this in the print for Ives' *Report* (cat. 27a), perhaps at the request of the geologist, Dr. Newberry, who probably did not consider Möllhausen's exaggeration amusing. The geologist described the canyon as not bearing "comparison, in point of magnitude, with several others traversed." Yet this site seemed to prove Newberry's theory of canyon formation along the Colorado River:

> The walls of Pyramid cañon are, for the most part, formed of massive granite, which exhibit the perpendicular faces two hundred feet or more in height. On either side of the stream, among the summits of the granite hills, full two hundred and fifty feet above its bed, are masses of stratified gravel, identical in composition, and continuous with the gravel terraces which border the river both above and below the cañon. Here, then, we have conclusive evidence that where now is Pyramid cañon was once an unbroken barrier, stretching across the course of the Colorado, and raising its waters to an elevation of at least two hundred and fifty feet above their present level. This barrier has been cut through by the action of the river, but these beds of gravel and sand remain as monuments of the time when the valley above Pyramid cañon, as well as that of the Mojaves below, were in great part covered by sheets of water.

> Similar deposits would doubtless have been found on the sides of the other cañons of the Colorado; but such accumulations have generally, from the greater height of the barrier, been less in quantity, longer exposed to atmospheric degradation, and, as a consequence, have disappeared.[8]

1 Ives, *Report*, part 1, p. 75. See USGS 1:250,000 map "Kingman"; USGS 15' series map "Davis Dam Quadrangle"; USGS 7.5' series "Davis Dam Quadrangle." This last map clearly shows Bull's Head Rock.

2 Ives called the Newberry Mountains the Dead Mountains. See cat. no. 30.

3 George A. Johnson, "The Steamer General Jesup," *Quarterly of the Society of California Pioneers* 9, no. 2 : 108-118.

4 Newberry in Ives, *Report*, part 3, p. 32.

5 Ives, *Report*, part 1, p. 75. Compare Ives' original map, plat 43, Egloffstein's printed map 1, and USGS 7.5' series "Davis Dam Quadrangle."

6 Möllhausen, *Reisen*, chapter 15. Pyramid Rock or Obelisk Mountain stands (or stood) somewhere near Katherine's Landing on Lake Mohave. It is now under the deep waters of the lake. In a 1990 visit we also found nothing resembling the curious formation in the distant right of Möllhausen's view. Granger wrote in *Arizona's Names*, p. 505: "Lt. Joseph Christmas Ives named this canyon in 1857 because near the rapids there was a natural pyramid nearly thirty feet tall. The location was known as Pyramid Rock, which now lies underwater." Barnes, *Arizona Place Names*, p. 51, cites a 1921 GLO map for the pre-dam location.

7 Ives, *Report*, part 1, pp. 89-90.

8 Newberry, in Ives, *Report*, part 3, pp. 32-33.

CAT. NO. 28

Balduin Möllhausen
Jessups Rapid (Deep Rapid)
Watercolor and gouache on paper
8⅜ x 11¼ in. (21.3 x 28.6 cm)
signed l.r.: "Möllhausen"
l.l. on paper mount: "37"
l.c. on paper mount: "Jessups rapid."
ACM 1988.1.35

The title of Möllhausen's watercolor, *Jessups Rapid*, refers to the point where Captain George A. Johnson's steamboat *General Jesup* was forced to turn back in January 1858. Ives and the *Explorer* reached these rapids in the late afternoon of February 18. The lieutenant wrote in his *Report* that above Pyramid Canyon,

> rapids were encountered in quick succession, and have been met with, at short intervals, up to camp, which is twenty miles from the head of the Mojave valley. Most of them have been ascended without difficulty. At one (Deep rapid) there was sufficient depth and channel unobstructed by rocks, but the rush of water was very strong. When we first heard its roar and saw the surging and foaming torrent we were startled, and a little apprehensive that we might have reached the head of navigation. There was less difficulty in making the ascent than had been anticipated. Not knowing what depth of water would be found, Captain Robinson had the boat lightened and Mr. Carroll put on a head of steam that made the stern wheel spin around like a top, and a line being taken out ahead, the summit of the rapid was quickly attained.[1]

Later, Ives added that Iretéba "had thought that the Deep rapid would put a stop to the steamboating, and since that has been passed entertains a higher opinion of the capabilities of our craft."[2]

In *Reisen*, Möllhausen wrote candidly that at first the force of water in this particular rapid forced them back downstream:

> Since we would have to take the usual precautions here, and since evening was almost upon us, our captain postponed any renewed attempts to the following day, and steered the boat over to the left bank. She was securely moored in calm water

behind a prominent rock while we selected a suitable place for our evening's encampment farther upstream. The journey today consisted of eleven miles, and we had reached the point where the steamboat *Jesup* had turned around a few weeks earlier. A little pyramid built of rubble with a stake on its top marks the spot we named "*Jesup*'s Halt."[3]

Cat No. 28

A comparison of Möllhausen's original watercolor with the print titled *Deep Rapid* in Ives' *Report* (cat. 28a), together with Möllhausen's journal and Ives' official report, gives further evidence of the intense rivalry between Lieutenant Ives and Captain Johnson.[4] It was not by chance that Ives failed to mention Johnson or the *General Jesup*. On plat 44 of the original map, Ives or Egloffstein designated a point on the river next to camp 48 with the words "Rapid" and "Jessup turned back," but on the official

28A
John James Young after Balduin Möllhausen, *Deep Rapid*, engraving, from Ives, *Report* (1861), part 1, p. 77, Amon Carter Museum Library

printed map, only the description "Deep Rapid" appears. Also, instead of using the title of Möllhausen's original watercolor, someone—probably on orders from Ives—changed the engraving's title to *Deep Rapid* (cat. 28a). The marker in the right of the watercolor was also omitted from the engraving. At far right in both images rests the bow of the moored *Explorer*. For whatever reason, the small fieldpiece is missing from the bow, even in the watercolor; perhaps the men had already unloaded it to lighten the boat and had dragged it to their encampment as protection against hostile Indians.

Möllhausen continued his frank description by admitting that they "all wanted to leave *Jesup's* Halt as far behind . . . as possible." This desire drove most of them to climb the nearby heights so they could overlook the river for a few miles. The view was "not very encouraging," however, since as far as they could see along the river, there were new rapids. Möllhausen then described how the captain and his disembarked crew guided the steamboat through the rapid early the next morning, with the aid of a long towrope. Afterwards, Möllhausen recorded: "We were reassured that we had actually pushed on somewhat farther than the *Jesup*. Even our crew expressed their satisfaction about it, and they worked twice as hard in portaging our cargo a few hundred feet upriver, where they loaded it on board again."[5]

The *Explorer* made only about three miles on February 19 before another rapid halted the day's progress. The crew spent almost the entire next day trying to get past this obstacle, and by dark, when they finally had the steamer above it, a fierce norther blew in. After losing a day to the sandstorm, the crew had to drag the steamboat through some shallows for most of February 22. Another sandstorm blew up the next day, and it was not until February 24 that the steamboat resumed the journey upriver. After reporting that they "ate, drank, breathed, and saw little but sand for twenty-four hours, and the gale was so violent that the Explorer was dragged from her anchorage and driven upon the rocks," Ives noted:

The late detentions have afforded Dr. Newberry and Mr. Egloffstein excellent opportunities to pursue their respective avocations. The doctor has had leisure to make a very full and perfect mineral collection, and become thoroughly conversant with the geological characteristics of the region.

Mr. Egloffstein has taken panoramic views of the river and the adjacent country, and has now completed a set that extends from Fort Yuma to the present camp. The ascent of a prominent peak on the opposite side of the river (Mount Newberry) has given him a view of the whole of the Black mountain range.[6]

At least two other sketches were made near this camp (no. 49). Möllhausen's sketchbook contains a view (cat. 28b), dated February 21, from a cave—perhaps the one where he and several other men sought shelter from blowing sand. Möllhausen described the cave in his journal entry for Feb. 21:

In the high bank a few paces from our camp, I discovered a cave which had been formed by water plunging down at that level from time to time. It was almost completely hidden behind roots and tendrils and promised some protection from the rampant storm. The tents which had been pitched in the

28B
Balduin Möllhausen, *View from Cave, February 21*, watercolor and graphite (no opaque white), from Sketchbook 3, Möllhausen Family, Bleicherode, Germany, photograph courtesy of Dr. Friedrich Schegk, Munich

sandy soil could not possibly have withstood the storm. Using an ax, I cleaned out the low brush from the cave, and left only enough roots and tendrils in place to provide a makeshift cover. My comrades helped me level the soil with a thick layer of willow branches, and so we soon created a relatively comfortable den which was spacious enough for our party of seven people.[7]

The view is the only surviving field sketch relating to the Ives expedition that contains watercolor, but since it has graphite underdrawing, the artist may have added this coloring later.[8] In the sketch, a bearded figure (perhaps a self-portrait?) seated on a campstool, watches an Indian at right, who in turn contemplates a mountain, probably Mount Newberry, in the distance.

The original sketch for the print titled *Hills of Tertiary Conglomerate and Modern Gravel, Camp 49* (cat. 28c) no longer exists. Möllhausen or Dr. Newberry may have made the lost field sketch

during their February 22 excursion together. One or both men may appear in the print, which shows one figure raising a geologist's hammer while another stands guard with a rifle.

Returning downriver the following month, the expedition camped on an island just above Jesup's Halt on March 19 and the next morning passed down the rapids "without incident," according to Möllhausen.[9] The former rapid is today completely inundated by the waters of Lake Mohave, at the northeastern base of the mountain that Ives and his men called "Mount Newberry" after their geologist.[10] The waters of Lake Mohave have also inundated the areas depicted in the two field sketches (cats. 28b and 28c), making identification of many of the topographical features quite difficult.

1 Ives, *Report*, part 1, p. 75-76.
2 Ives, *Report*, part 1, p. 77.
3 Möllhausen, *Reisen*, chapter 15.
4 For a discussion of this rivalry, see Arthur Woodward, *Feud on the Colorado*.
5 Möllhausen, *Reisen*, chapter 15.
6 Ives, *Report*, part 1, p. 76-78; compare this with Möllhausen, *Reisen*, chapter 16. Egloffstein's panoramic view, titled *Colorado River, Cottonwood Valley, From Pyramid Cañon to Painted Cañon*, is reproduced in Ives' *Report*, part 1, opp. p. 74; compare it with USGS 1:250,000 map "Kingman" and with Ives' maps.
7 Möllhausen, *Reisen*, chapter 16.
8 In *Reisen*, chapter 15, Möllhausen describes painting Indian bodies and faces with watercolors, so he clearly had such materials with him on the expedition.
9 Möllhausen, *Reisen*, chapter 19.
10 Compare Egloffstein's maps with USGS 1:250,000 map "Kingman."

28c
John James Young after Balduin Möllhausen or John Strong Newberry, *Hills of Tertiary Conglomerate and Modern Gravel, Camp 49*, engraving, from Ives, *Report* (1861), part 3, p. 34, Amon Carter Museum Library

CAT. NO. 29

Balduin Möllhausen
Cottonwood Valley
Watercolor and gouache on paper
8¼ x 11¼ in. (21.0 x 28.6 cm)
signed l.r.: "Möllhausen"
l.l. on paper mount: "35"
l.c. on paper mount: "Cottonwood valley"
m.l. on paper mount: "M. Davis"
ACM 1988.1.33

29A
John James Young after Balduin Möllhausen, *Cottonwood Valley*, engraving,
from Ives, *Report* (1861), part 1, p. 79, Amon Carter Museum Library

Cat. No. 29

The watercolor titled *Cottonwood Valley* probably depicts an area near the expedition's camp (no. 52) for the evening of February 26. Möllhausen recorded that this campsite (today under the waters of Lake Mohave) was "on the left bank in a magnificent grove of cottonwood trees." For the entire day, as the steamboat approached the area, they had "seen the trees shimmering in a most inviting way" in the distance. They named the valley "in honor of the cottonwood trees,"[1] which according to Ives, were "of a larger growth than any seen before." The lieutenant wrote that the valley was

> only five or six miles in length and completely hemmed in by wild-looking mountains. The belt of bottom land is narrow, and dotted with graceful clusters of stately cottonwood in full and brilliant leaf. The river flows sometimes through green meadows, bordered with purple and gold rushes, and then between high banks, where rich masses of foliage overhang the stream, and afford a cool and inviting shade. From the edges of this garden-like precinct sterile slopes extend to the bases of the surrounding mountain chains.[2]

Several tall cottonwoods rise from thick vegetation at left in the watercolor and engraving (cat. 29a). In the distance, more cottonwoods line the river's edge while the bare mountains mark the limits of the valley. An inscription on the watercolor's margin denotes the mountain at far right as Mount Davis (see cat. 31c).

Similar trees appear in a field sketch dated February 25, also taken in the Cottonwood Valley (cat. 29b). Möllhausen's diary comments on the "magnificent" weather that day and the sight of groves of trees:

> Even though they were scattered very far apart, I can hardly describe how wonderful was the sight of fresh, living vegetation in the midst of a stark, rocky desert. We contemplated and admired trees which would scarcely be noticed in more favorable regions. When we looked at the masses of volcanic rock which surrounded us, we quickly turned again to the fresh lovely green spring colors of lonely [isolated] trees and bushes.[3]

In the sketch, Möllhausen emphasized the vegetation along the river at the expense of landforms, which he rendered much more faintly. The date suggests that the area depicted (between camps 50 and 51) is also under Lake Mohave today.[4]

The explorers observed that only a few scattered Mohave families inhabited the valley, and Ives recorded that "we saw no fields under cultivation, and the residents brought neither corn

29B
Balduin Möllhausen, *Colorado River Landscape, February 25*
["Flußlandschaft"], graphite on paper, from Sketchbook 3,
Möllhausen Family, Bleicherode, Germany, photograph
courtesy of Dr. Friedrich Schegk, Munich

and moved slowly backwards with the current until they stood across from a suitable depression in the bank. The two fishermen who carried the ends circled around carefully and approached the bank. Together with the others they pulled the net and its contents out of the water. Small fish easily slipped through the wide mesh, but they caught enough large ones to delight our entire party with a hearty meal. This was of great importance since the constant consumption of salt pork and the complete lack of fruits and vegetables caused us to fear an outbreak of scurvy.[7]

Around this time, Möllhausen also recorded that the expedition met a number of natives who, according to their Mohave guide, were "Chemehuevis and bad Indians." The German wrote that

they did not appear to have anything in common with the tribe of Chemehuevis living to the south except the name. They were such an unclean society, the likes of which we had not seen on the Colorado except among the Apaches. They had slender, but very haggard builds, but lacked the natural grace that characterizes the Mohaves, Yumas, and Southern Chemehuevis. Their expressions were not as open, but instead aroused gloom and mistrust.[8]

A field sketch dated February 25 (cat. 29c) shows two Indians who fit this description. Both wear breechclouts, and the standing figure carries a bow and also has a tunic or cloak of what may be an animal skin.

Möllhausen wrote that, on February 27, "with some regret . . . we left that garden-like grove of trees . . . for from now on we would encounter only barren rock masses and arid stretches of sand." They made only three miles that day, but that was "far enough to lead us into the desert where there was not even enough wood to fire our boiler."[9]

Over two weeks later, on March 15, Lieutenant Ives and his men returned to this area after reaching the head of navigation farther upriver and remained there for several days to enjoy the shade trees again and to trade with the Mohaves for food. Here they also encountered a lone Mormon and finally met up with G. H. Peacock, Lieutenant Tipton, and the mule train from Fort Yuma, which arrived with much-needed food and supplies on

nor beans to trade."[5] Möllhausen, however, noted: "like those living to the south they had a friendly disposition toward the whites. They showed this friendly disposition by providing us with fish. Even when we were deep in the mountains farther upstream they followed us with their nets and received many a string of glass beads from us for the fish they caught."[6]

Some of these Mohaves appear in the right foreground of Möllhausen's watercolor, extending their seines or drag nets into the deeper part of the channel. The artist had watched another group of Mohaves fishing on January 31, below Corner Rock, and had noted that their

net was made of coarse mesh pieced together from fine, but very strong, threads of inner bark fiber. The net was about four feet high and about thirty feet long. Long stakes every four feet held the net upright in the water and secured it to the ground. This was the only gear our fishermen had. Holding the net stretched taut, five or six people waded into the river

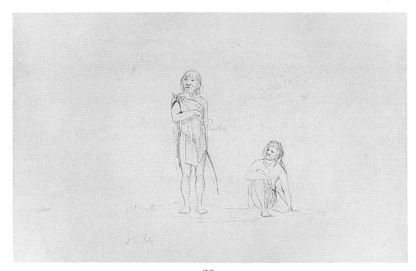

29c
Balduin Möllhausen, *Two Chemehuevis, February 25*
[*"Zwei Indianer, 25. Febr."*], graphite on paper, from Sketchbook 3,
Möllhausen Family, Bleicherode, Germany, photograph courtesy
of Dr. Friedrich Schegk, Munich

March 18. Möllhausen commented that, had they postponed
conferring a name on the valley until their second visit, "perhaps
we would have named it 'Valley of Plenty.'"[10]

1 Möllhausen, *Reisen*, chapter 17. Compare USGS 7.5' series quadrangle "Spirit
Mtn. NW." and Egloffstein's map 1.
2 Ives, *Report*, part 1, p. 78. Newberry describes Cottonwood Valley in three pages
(pp. 33-35).
3 Möllhausen, *Reisen*, chapter 16.
4 Compare Egloffstein's printed map 1 and USGS 1:250,000 map, "Kingman."
5 Ives, *Report*, part 1, p. 79.
6 Möllhausen, *Reisen*, chapter 17.
7 Möllhausen, *Reisen*, chapter 11. Modern diagrams and photographs of traditional Mohave fishing equipment appear in Kenneth M. Stewart, "Mohave,"
in Sturtevant and Ortiz, eds., *Handbook of North American Indians*, vol. 10,
Southwest, p. 61. Möllhausen did not show or mention that they used rock
weights to hold down the bottom edge of the seine.
8 Möllhausen, *Reisen*, chapter 16.
9 Möllhausen, *Reisen*, chapter 17.
10 Ibid.

CAT. NO. 30

Balduin Möllhausen
Painted Cañon
Watercolor and gouache on paper
8⅜ x 11⅛ in. (21.3 x 28.3 cm)
signed l.r.: "Möllhausen"
l.l. on paper mount: "No 40"
l.r. on paper mount: "Painted Cañon"
ACM 1988.1.38

Around noon on February 27,
at the head of Cottonwood Valley,
the *Explorer* reached another canyon,
formed, according to Lieutenant
Ives, "by the passage of the river
through a spur that connects the
Black and Dead mountain ranges."
The lieutenant described it as

Cat. No. 30

only two or three miles in extent,
and the sides were of moderate height, but the gorgeous
contrast and intensity of color exhibited upon the rocks
exceeded in beauty anything that had been witnessed of a
similar character. Various and vivid tints of blue, brown, white,
purple, and crimson, were blended with exquisite shading
upon the gateways and inner walls, producing effects so novel
and surprising as to make the cañon, in some respects, the
most picturesque and striking of any of these wonderful
mountain passes."[1]

Möllhausen noted: "The walls rose straight up . . . alternating
masses of blue-black lava and trachyte, vividly colored columns of
porphyry, gray conglomerate, and colorful sandstone. For a half a
mile, the entire length of the canyon, we had a wonderful array of
colors before our eyes, the likes of which we had seen only at the
Riverside Mountains. So we named this place Painted Canyon."[2]

The twisted rock formations in the center of Möllhausen's
Painted Cañon watercolor are geologically implausible, and Dr.
Newberry may have requested the engraver to correct them in
the print for Ives' *Report* (cat. 30a). However, besides straightening
the walls of the canyon, the engraver increased their height and

magnitude and that of the mountains beyond, and reduced the size of the steamboat. The resulting exaggerated scale in the print renders the topography almost totally unrecognizable.

Most of Painted Canyon is today under the waters of Lake Mohave, and only a few traces of its colorful formations remain visible. Its southern entrance was near present-day Cottonwood Cove or Cottonwood Landing on the Nevada side of the lake. The northern end of Ives' "Dead mountain range" is now called the Eldorado Mountains.[3]

---

1 Ives, *Report*, part 1, p. 79.
2 Möllhausen, *Reisen*, chapter 17. Dr. Newberry described the canyon in Ives, *Report*, part 3, p. 35.
3 USGS 7.5' series "Mt. Davis" and "Spirit Mtn. NW" quadrangles; Granger, *Arizona's Names*, p. 457; and USGS 1:250,000 map "Kingman."

30A
John James Young after Balduin Möllhausen, *Painted Cañon*, engraving, from Ives, *Report* (1861), part 1, p. 80, Amon Carter Museum Library

CAT. NO. 31

Balduin Möllhausen
Dead Mountain. View From Round Island
(Dead Mountain, Mojave Valley)
Watercolor and gouache on paper
8½ x 11⅛ in. (21.6 x 28.3 cm)
signed l.r.: "Möllhausen"
l.l. on paper mount: "41"
l.c. on paper mount: "Dead M. view from Round Island."
ACM 1988.1.39

Cat. No. 31

For several nights in late February and early March, the expedition camped (no. 53) at a "Round Island" near the base of Mount Davis, and from the site Möllhausen sketched a mountain to the south that they had already passed. One or more of these sketches aided the artist in his later watercolor of this peak, which Lieutenant Ives named Dead Mountain. Modern maps denote this mountain (elevation 5,639 feet) as Spirit Mountain, located in the Newberry Mountains of Nevada, just within the western boundary of Lake Mead National Recreation Area. (This mountain is sometimes confused with Mount Newberry, a smaller mountain also named by Ives in the same range.[1])

In his entry for February 27, Möllhausen recorded:

On this day we had a beautiful view of the mountain behind us. A week before, near its base, we had passed beyond *Jessup*'s Halt. Seen from the north, the mountain provided a vista of picturesque lines and forms. Iretéba, who had noticed that we had turned our attention to the mighty mountain, told us that this mountain had special significance for the natives of the Colorado Valley, for it was the residence of all Mohaves who had gone to the great beyond. His brother, who had been slain by the Coco-Maricopas, already resided there, and some day he himself would also move there. Since the good hearted

Indian believed that his brother had gone to that place, I concluded that only good people were sent to the mountain. In any event, this was the only instance I knew of where one of the natives there made a voluntary allusion to the afterlife of the soul.[2]

Lieutenant Ives noted that, according to Iretéba, "should any one dare to visit it he would be instantly struck dead."[3]

Many other sources from both the nineteenth and twentieth centuries record sacred myths, legends, and stories connected with this mountain. The Yuma or Quechan call it "Avikwamé"; it figures prominently in their creation stories and those of other Colorado River tribes, including the Mohave.[4] Even Dr. Newberry, whose geological report is sparse on nonscientific matters, did not fail to notice the spiritual significance of Dead Mountain for the natives.[5]

The natives' reverence for Avikwamé and its departed spirits, the expedition's position on an island surrounded by a desolate landscape, and the pervading fear of hostile Indians created a real-life situation almost straight out of James Fenimore Cooper's *The Deerslayer*. No doubt these associations stirred the explorers' own romantic notions of the beautiful, the picturesque, and the sublime. Möllhausen's narrative is full of such words, and at Round Island he vividly described listening at night to the calls of wild animals in their otherwise silent surroundings and watching the moon rise over Mount Davis.

It is easy to understand how John James Young or the printmakers back in Washington, D.C., who heard such stories, may have decided that Möllhausen's original watercolor was not suggestive enough. The transfer lithograph of Dead Mountain in Ives' *Report* (fig. 31a) has little in common with Möllhausen's watercolor. Whereas the watercolor depicts the specific shape of the sacred mountain in daylight, the lithograph shows a generalized and uniformly shaped mound in gloomy, romantic moonlight. Rocks and gravel litter the riverbed in the watercolor, but in the print, a smooth reflection of moonlight outlines the dark silhouettes of two Indians on a rocky prominence, who appear to gaze at the mountain and moon with awe and reverence. In the watercolor original, this same prominence is a wooded projection of Round Island where one of the explorers (perhaps the artist?) sits next to his gun and seems engaged in writing or sketching. Young and the printmakers may also have been inspired by Möllhausen's field

31A
John James Young after Balduin Möllhausen, *Dead Mountain, Mojave Valley*, toned lithograph (black with gray ruled screen), from Ives, *Report* (1861), part 1, opp. p. 75, Amon Carter Museum Library

sketch of a view from a cave (see cat. 28b), which shows an Indian similarly contemplating a mountain from across a rapid.

While at Round Island on March 1, Möllhausen made a study sketch of a stump or pile of driftwood (cat. 31b), similar but not identical to the one in the foreground of his watercolor titled *Dead Mountain*. His careful observation of nature is evident in both the sketch and his written description of Round Island:

> Its mounds of sand and piles of gravel were covered with large masses of driftwood. Small stands of willowbrush grew here and there in the damper places. The crowns of three or four misshapen cottonwood trees leaned toward the rushing torrent, but otherwise the island and the surrounding terrain could be called an arid dreadful wilderness.[6]

Adding to the sense of dread, Möllhausen recorded that Iretéba and Mariano warned the whites that they were now in Paiute country and should be on their guard. The men discovered signs that natives had occupied the island several months earlier.[7]

31B
Balduin Möllhausen, *Driftwood ["Baumstumpf"]*, graphite on paper,
from Sketchbook 3, Möllhausen Family, Bleicherode, Germany,
photograph courtesy of Dr. Friedrich Schegk, Munich

(Iretéba's precautions were apparently well-founded, for six years later on "Cottonwood Island"—probably Round Island—Paiutes and a militant faction of Mohaves took Iretéba prisoner and, before releasing him, humiliated him by stripping him of the major general's uniform that the whites had given him as a gift.[8])

Round Island lay at the base of Mount Davis, where, according to Ives, "the river divides and forms a round island of considerable extent, at the foot of which is a rapid which has created some trouble and detention."[9] The expedition's maps show rapids at both ends of the island, and this is confirmed by Möllhausen, who described "the roaring of powerful rapids" and wrote of lightening the vessel to get it over a shallow bar of gravel and rocks.[10] He also noted their arrival, on the evening of February 27, at the base of a "small but unusual" mountain that the lieutenant named for Secretary of War Jefferson Davis. "Several times I tried to convince Lieutenant Ives to use Indian names," Möllhausen complained, "but he said he was firmly opposed to pagan nomenclature that no one would be able to spell or pronounce. Even when I wanted to bestow the names Iretéba and Kairook to mountains as a memorial

in honor of tribes which would soon disappear, he raised objections which do not make any more sense now than they did then."[11]

Ives described Mount Davis, "[j]ust above the Painted cañon," as "a symetrical [*sic*] and prominent peak . . . which presents the most conspicuous landmark north of the Dead mountain."[12] Located in Mohave County, Arizona, Mount Davis continues to bear the name of the former Secretary of War, who was also a Mississippi congressman, Mexican War hero, and eventually president of the Confederacy. (During the Civil War, Ives would serve as President Davis' personal aide.[13]) Möllhausen would no doubt be somewhat mollified to learn that on the Clark County, Nevada, side of the river, some of the Eldorado Mountains now bear the name Ireteba Peaks.[14]

Möllhausen noted that on the west side of Mount Davis, "lava-like slopes extended to the river." Likewise, Dr. Newberry speculated that all of the nearby mountains had "the appearance of being upheaved fragments of an ancient lava-plain." He wrote that he was "rather disposed to regard them as only the remnants of broad rounded masses of erupted rock, each marking the sight [*sic*] of a volcanic vent, worn into angular forms which they exhibit by the erosion to which they have been exposed."[15]

The print of Mount Davis in Ives' *Report* (cat. 31c) was probably loosely based on a Möllhausen sketch, now lost. This print does not accurately convey the mountain's peculiar topographical characteristics but exaggerates the steepness of the slopes and the sharpness of the peak.[16] However, it does show the *Explorer's* crew members hauling the steamboat over rapids—an activity recorded by all sources. Another print (cat. 31d), after a panoramic view that Egloffstein likely made from Mount Davis, shows two men rowing the skiff out ahead of the *Explorer*, which probably was grounded again in the shallows.

On March 2, 1858, as the expedition made its way upriver, Möllhausen described "colorful low rock masses" that "extended up to the river on both sides, alternating with remnants of the gravel plain resembling walls. As often and as long as we had observed them, they did not show the slightest change in their outer character." Seven miles north of Round Island, the *Explorer* again encountered a gravel bar at a bend in the river, and "the rest of the day passed with unloading, winding, and then loading up again."[17] Since Mount Davis remained in view at this time, Möllhausen might have drawn his lost sketch from here, although

31c
John James Young after Balduin Möllhausen, *Mount Davis*, engraving, from Ives, *Report* (1861), part 1, p. 81, Amon Carter Museum Library

the inexactness of the print makes it impossible to determine exactly where or when.

On March 2, Möllhausen noted, they "landed on the left bank on a narrow ribbon of sand which extended along the base of the gravel hills (up to 200 feet in height)." He and several others climbed the heights above this camp (no. 54) and followed an old Indian path "to a place on the heights which offered a broad vista of the jumbled mountain ranges." He recorded that "the loneliness which reigns there is indescribable. Everything was dead and barren in all directions, including the trachyte mountains and their dolomite outcroppings and the gray ascending plain with its arroyos and deep embrasures."[18]

31d
After F. W. von Egloffstein, Detail of *Colorado River, Davis Valley, From Painted Cañon to Black Cañon*, lithograph, from Ives, *Report* (1861), part 1, opp. p. 84, Amon Carter Museum Library

The next day, they repeatedly saw the remarkable gravel embankments right at the water's edge. At times the embankments reminded us of the gray ruins of medieval buildings [cat. 31e]. By contrast, magnificent volcanic mountains towered up in the background, embellished with picturesque lines and beautiful colors, but the desert's changing images lay reflected on the stream's surface as well as in reality. Sublime nature lay before us and moved our souls.[19]

For the most part, however, both Möllhausen's and Ives' journal entries for that day were unusually reticent. Möllhausen observed that they "made five miles and . . . found little opportunity to enrich our diaries with notes and observations."[20]

1 Compare Ives' original and printed maps with USGS 7.5' series "Spirit Mtn. Quadrangle." Jack D. Forbes, *Warriors of the Colorado: The Yumas of the Quechan Nation and Their Neighbors* (Norman: University of Oklahoma Press, 1965), p. 22, identifies the sacred mountain of the Colorado tribes, "Avikwamé," as "Newberry Mountain, north of Needles." However, it is obvious that Ives and his men believed that Iretéba was referring to Dead Mountain, which is the taller of the two. Dead Mountain also appears in the distance of Möllhausen's watercolor of the Head of Navigation (see cat. no. 33).

2 Möllhausen, *Reisen*, chapter 17.

3 Ives, *Report*, part 1, p. 75.

4 Forbes, *Warriors of the Colorado*, pp. 4, 22, 25, 27-30, 44, 63, 65, 344. Today only a portion of the mountains on the west side of the Colorado River are still called "Dead Mountains." These lie opposite the Fort Mohave Indian Reservation in San Bernardino County, California.

5 See Ives, *Report*, part 3, p. 32.

6 Möllhausen, *Reisen*, chapter 17.

7 Ibid.

8 See Arthur Woodward, "Irataba—Chief of the Mohave," *Plateau, Museum of Northern Arizona* 25 (January 1953): 64-66, including footnotes.

9 Ives, *Report*, part 1, p. 79.

10 Möllhausen, *Reisen*, chapter 17. See Egloffstein's printed map 1 and original map, plats 46 and 47 and his map, camp 53, March 1.

11 Möllhausen, *Reisen*, chapter 17.

12 Ives, *Report*, part 1, p. 79.

13 For more on Davis, see *Dictionary of American Biography*, vol. 3, part 1, pp. 123-131, and Hudson Strode, *Jefferson Davis*, 3 vols. (New York: Harcourt, Brace and World, Inc., 1964).

14 See USGS 7.5' series "Mt. Davis Quadrangle"; Granger, *Arizona's Names*, p. 197; and USGS 1:250,000 map "Kingman." Walter R. Averett, *Directory of Southern*

Nevada Place Names, rev. ed. (Las Vegas: the author, 1963), p. 55, does not record who named the Ireteba Peaks. Perhaps some sharp cartographer had read Möllhausen's criticism.

15 Möllhausen, *Reisen*, chapter 17, and Newberry in Ives, *Report*, part 3, pp. 35-36.

16 Mount Davis also appeared in Möllhausen's watercolor view *Cottonwood Valley* (cat. no. 29) and in Egloffstein's panoramic view titled *Colorado River, Cottonwood Valley, From Pyramid Cañon to Painted Cañon*, in Ives, *Report* (1861), part 1, opp. p. 74.

17 Möllhausen, *Reisen*, chapter 17.

18 Ibid.; also see Newberry in Ives, *Report*, part 3, pp. 36-37.

19 Möllhausen, *Reisen*, chapter 17. Around this time, in Ives, *Report*, part 3, pp. 38-39, Dr. Newberry took note of Elephant Hill, visible in early March from camps 55, 56, and 57 and "scarcely two miles from the stupendous gateway which forms the southern entrance to the Black cañon." He made a geological cross section of it and also reported finding "a very large and perfect tooth of *Elephas primigenius* in the bed of coarse gravel and boulders which forms its base." Ives marked it Elephant Butte on his original map.

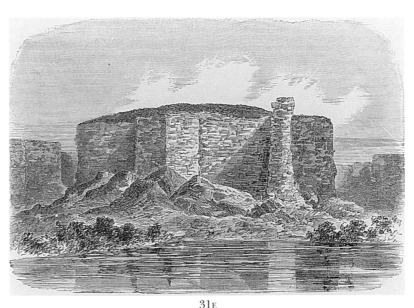

31E
After Balduin Möllhausen, F. W. von Egloffstein, or J. S. Newberry, *Castle-like Conglomerate Bluff*, engraving, from Ives, *Report* (1861), part 3, p. 36, Amon Carter Museum Library

CAT. No. 32

Balduin Möllhausen
***Twelve Miles South [sic., North] of Round Island
(Gravel Bluffs South of Black Mountains)***
Watercolor and gouache on paper
8⅜ x 11¼ in. (21.3 x 28.6 cm)
signed l.r.: "Möllhausen"
l.l. on paper mount: "39"
l.c. on paper mount: "12 M. south of Round Island"
ACM 1988.1.37

Cat. No. 32

The watercolor inscribed "12 M. south of Round Island" served as the model for the print *Gravel Bluffs South of Black Mountains* (cat. 32a). The engraver faithfully reproduced many of Möllhausen's topographical details, although he compressed the composition. The print reduces the number of figures on the Mohave raft in the foreground from five to four and omits some of their baggage, including a large basket or ceramic pot. Möllhausen's watercolor shows the raft to be constructed of bundles of reeds or sticks, but the engraver also omitted these details.

The inscription on the watercolor suggests that Möllhausen made the original sketch in the Cottonwood Valley, in the vicinity of camps 50 and 51, perhaps on February 24 or 25. However, Ives inserted the print in a later portion of his *Report*, suggesting that it may depict a view farther upriver, looking north toward the southern entrance of Black Canyon; a recent examination of the site tends to support Ives' placement. The tall mesa in the center distance of the watercolor might be Malpais Flattop Mesa, on the Arizona side of the river in the Lake Meade National Recreation Area—about twelve miles *north* of Round Island, near modern Nelson's Landing at the mouth of Eldorado Canyon, in Clarke County, Nevada.[1] In that case, a date of March 4 or 5 (when they made camps 55 and 56) would be more plausible for the original sketch.

In camp at Round Island on March 1, Lieutenant Ives may have been referring to the canyon walls in the left distance of the watercolor when he asserted:

We have now entered a region that has never, as far as any records show, been visited by whites, and are approaching a locality where it is supposed that the famous "Big Cañon" of the Colorado commences; every point of the view is scanned with eager interest. We can distinctly see to the north the steep wall of one side of the gorge where the Colorado breaks through the Black mountains. Whether this is the "Big Cañon" or not it is certainly of far grander proportions than any which we have thus far traversed.[2]

Möllhausen mentioned sketching the southern portal of Black Canyon. As they approached on March 6, he sat on the platform of the *Explorer* and "looked at the terrible rock masses which rose

32A
John James Young after Balduin Möllhausen, *Gravel Bluffs South of Black Mountains*, engraving, from Ives, *Report* (1861), part 1, p. 82, Amon Carter Museum Library

straight up a thousand feet and delineated the portals of the much discussed and decried canyon." When they landed to take on wood for the boiler, he "used the time to sketch the rock portal which lay before us several hundred paces away. The first rock walls rose no more than three hundred feet high, behind them new overhanging masses of black volcanic rock jutted up."[3] This sketch apparently no longer exists; the view in the watercolor probably depicts the same area from a point farther down the river.

---

1   Compare USGS 1:250,000 map "Kingman" and 7.5' series "Fire Mountain" quadrangles with Ives' original map, plats 48,49,50 and printed map 1.

2   Ives, *Report*, part 1, p. 79. Also see Egloffstein's map. Barnes notes in *Arizona Place Names*, p. 52 and 55:

> Black Range, Mohave Co., U.S.G.S. Map, 1923. Often called Black mountains. An extension to the south of Black Canyon range from T. 23 N., R. 20 W. So called by Ives who writes: "The range east of the Mohave Valley we call the Black Mountains.... Where the river breaks through the chain there is doubtless a stupendous canyon." Coues writes: "The earliest name for this range I know is the Black Range of Ives' *Report*. In my time (1865) it was called the Sacramento Range from the name of the valley to the east. Garces calls it Sierra de Santiago or St. James Range." The Sacramento range does not appear on any map. Lieut. Mallory, however, according to Hinton called the southern end of this range Blue Ridge mountains. The northern end he called Black Canyon range. Both are on early maps. P.O. est. as Blue Ridge April 20, 1917.

3   Möllhausen, *Reisen*, chapter 17.

Cat. No. 33

Balduin Möllhausen
Head of the Navigation (Mouth of the Black Cañon)
Watercolor and gouache on paper
8¼ x 11¼ in. (21.3 x 28.6 cm)
signed l.r.: "Möllhausen"
l.l. on paper mount: "42"
l.c. on paper mount: "Head of the navigation"
ACM 1988.1.40

Both Möllhausen and Lieutenant Ives recounted how the *Explorer* dramatically reached the limit of its journey up the Colorado River. On March 6, they could see the southern portal of Black Canyon directly in front. According to Ives:

Cat. No. 33

> A rapid, a hundred yards below the mouth of the cañon, created a short detention, and a strong head of steam was put on to make the ascent. After passing the crest the current became slack, the soundings were unusually favorable, and we were shooting swiftly past the entrance, eagerly gazing into the mysterious depths beyond, when the Explorer, with a stunning crash, brought up abruptly and instantaneously against a sunken rock. For a second the impression was that the cañon had fallen in. The concussion was so violent that the men near the bow were thrown overboard; the doctor, Mr. Mollhausen, and myself, having been seated in front of the upper deck, were precipitated head foremost into the bottom of the boat; the fireman, who was pitching a log into the fire, went half-way in with it; the boiler was thrown out of place; the steam pipe doubled up; the wheel-house torn away; and it was expected that the boat would fill and sink instantly by all but Mr. Carroll, who was looking for an explosion from the injured steam pipes. Finding, after a few moments had passed, that she still floated, Captain Robinson had a line taken into the skiff, and the steamer was towed alongside a gravelly spit a little below.[1]

No one had been seriously injured or even lost his journals or sketches in the accident.[2] Damage to the steamboat was, in the words of Ives, "such as a day or two of labor could repair." On March 8 he reported:

> Nearly three days have elapsed since the accident, and everything is restored to its former condition. I have thought it would be imprudent, after this experience of sunken rocks, to attempt the passage of the cañon without making a preliminary reconnaissance in the skiff. A second escape of the boat, in the event of a similar encounter with a rock, would be too much to hope for; and should she be sunk in the cañon, and there be nothing to swim to but perpendicular walls five hundred or a thousand feet high, the individuals on board would be likely to share the fate of the steamer.[3]

Möllhausen described the area that they all now realized was probably the head of navigation:

> Our surroundings consisted of rocks ranging from three hundred to a thousand feet high which were primarily volcanic. These perpendicular walls towered up high out of the river forming the left bank, and revealing enormous masses of trap and trachyte. By contrast, brightly colored porphyry rocks and greenish shimmering obsidian were arranged upon one another on the right bank.[4]

Dr. Newberry wrote enthusiastically that "probably nowhere in the world is there a finer display of rocks of volcanic origin than may be seen about the southern entrance to the cañon. The beetling crags which form its massive portals are composed of dark-brown porphyry of hardest and most resistant character."[5]

Early on the morning of March 8, Möllhausen hiked about half a mile to a projection above their camp (no. 57). He recorded: "I sat down on the last corner where I could overlook the southern portal of the canyon in order to make a sketch of the head of navigation. This very significant spot which lay before me was certainly a beautiful spectacle." There,

> [t]he river reflected the sun's splendor and was framed by the pale green ribbons of willows as it wound into the high rocks which framed the picture. In the far distance Dead Mountain rose up. All of its peaks and towers were enveloped in a

vaporous blue, and right beneath it the little *Explorer* lay in a backwater. The *Explorer* disappeared in its surroundings, so to speak, and seemed like a little speck when contrasted to such overpowering works of nature.[6]

In his watercolor, Möllhausen accurately depicted the topography of this site in the present-day Lake Mead National Recreation Area, just south of Fenlon Bend and some three miles north of Nelson's Landing, near Squaw Peaks. The Arizona side of the towering southern portal of the canyon looms at left, while in the distance to the south, Dead Mountain rises on the Nevada side. The "little speck" of the *Explorer* stands in the distance near the right bank. A recent photograph of the site (cat. 33c) indicates that little except the water level has changed in the intervening years.[7] The watercolor accurately delineates the characteristic features of Dead Mountain but brings it into closer focus. The geological details, however, are less exact and even inaccurate, suggesting that the artist's original sketch only recorded the

33A
John James Young after Balduin Möllhausen, *Mouth of Black Cañon*, engraving, from Ives, *Report* (1861), part 1, p. 83, Amon Carter Museum Library

33ʙ
After Balduin Möllhausen, *Ende der Schiffbarkeit des Rio Colorado.*
(Aussicht aus dem Black Cañon.), toned lithograph, from Möllhausen,
Reisen (Leipzig: Otto Purfürst, 1861), vol. 1, opp. p. 374

33ᴄ
The Head of Navigation, Black Canyon, March 1990

topographical outline of the formations, and that in later preparing the finished watercolor, he had to guess at their structure.

Prints reproduce this subject with a few alterations. The engraving for Ives' *Report* (cat. 33a) exaggerates the verticality and regularity of the canyon walls at left and right and replaces the two Indian figures in the watercolor with two smaller figures on a rocky projection in the center foreground. The engraver transformed the tiny *Explorer* into a two-masted craft, and, even more noticeably, brought the distant mountains still farther forward. The print for *Reisen* (cat. 33b) is more accurate topographically but introduces aquatic birds (perhaps herons) in the foreground.

Möllhausen and most of the men remained at the Head of Navigation for an entire week (March 6-13), expecting Lieutenant Tipton to arrive at any time with a train of packmules bringing supplies from Fort Yuma. The expedition members camped just below the point where the steamboat accident had occurred,

> on a sandy expanse which met the jumbled rocks of the right bank at an angle. It was a dreary stopping place which fate had forced upon us. Our steps were hindered either by deep sand, or by sharp volcanic rubble, and thorny mesquite trees grew where solid ground would have made it relatively easy to move about. There was absolutely no firewood, and we said it was fortunate that we had taken on a supply of it before we entered the canyon, for it now proved to be very useful. For the time being we were cut off from further travel, so we had to satisfy ourselves with what we had, and hope of better times to come.[8]

The camp was also "in an extremely vulnerable position" if a few good Mormon sharpshooters and their Paiute allies wished to "use the notches and clefts in the nearby rocks as fortresses."

Möllhausen recorded part of this camp in a field sketch dated March 9, depicting a tent with folding table and camp stools in front of a bluff (cat. 33d). Another field sketch (cat. 33e), near it in the artist's sketchbook, shows one Indian seated and another lying by a campfire and may also have been made at this campsite. A metal cooking pot on the fire in front of the natives suggests that they belong with the expedition. They both have agreeable expressions on their faces and long hair cropped above their eyebrows, and at least one of them draws a blanket over his shoulders. Möllhausen noted that the Mohave guides Iretéba and Navarupa were at this campsite, and although many of the men

33D

Balduin Möllhausen, *Tent with Equipment and Bluff in the Background*
[Zelt mit Ausrüstungsgegenständen und Felsen im Hintergrund],
graphite on paper, Sketchbook 3, Möllhausen Family, Bleicherode,
Germany, photograph courtesy of Dr. Friedrich Schegk, Munich

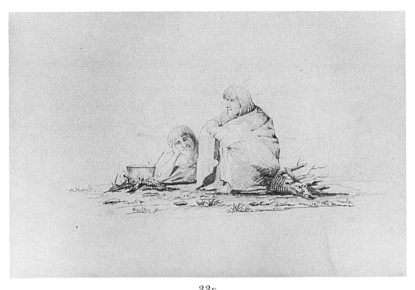

33E

Balduin Möllhausen, *Two Indians by a Fire (Iretéba and Navarupa?)*
[Zwei Indianer an Feuerstelle], graphite on paper, probably between
March 4 and 9, 1858, Sketchbook 3, Möllhausen Family, Bleicherode,
Germany, photograph courtesy of Dr. Friedrich Schegk, Munich

were sick from lack of adequate rations, the two Indians supplemented their diet by catching, roasting, and eating lizards and "meaty" salamanders, which they first offered to Möllhausen for his collection. "During the entire expedition our Indians were generally the healthiest people on the expedition, for while all of us became more or less ill, this life style seemed to suit them ideally, since they did not deviate from their own customs and habits."[9]

Before sunrise on March 9, Lieutenant Ives, having waited for Lieutenant Tipton and the pack train as long as he dared, set out from this base camp in the skiff to explore Black Canyon. Only Captain Robinson and the *Explorer*'s mate accompanied him. Quickly passing what Ives called "the outworks of the range," they soon found themselves in a canyon whose walls were "perpendicular, and more than double the height of those in the Mojave mountains, rising, in many places, sheer from the water, for over a thousand feet." For two days and part of another morning they pulled sculls and used a blanket as a sail to power them upriver against the strong current. Frequently they had to get out of the boat and haul it over rapids. Ives wrote:

The cañon continued increasing in size and magnificence. No description can convey an idea of the varied and majestic grandeur of this peerless waterway. Wherever the river makes a turn the entire panorama changes, and one startling novelty after another appears and disappears with bewildering rapidity. Stately façades, august cathedrals, amphitheatres, rotundas, castellated walls, and rows of time-stained ruins, surmounted by every form of tower, minaret, dome, and spire, have been moulded from the cyclopean masses of rock that form the mighty defile. The solitude, the stillness, the subdued light, and the vastness of every surrounding object, produce an impression of awe that ultimately becomes almost painful.[10]

Somewhere along the route, Lieutenant Ives made a rough sketch of Black Canyon, which Egloffstein later reinterpreted. Both

33ꜰ
F. W. von Egloffstein after Joseph C. Ives, *Black Cañon*, transfer lithograph (from an engraving?), from Ives, *Report* (1861), part 1, opp. p. 80, Amon Carter Museum Library

works are known today only through a transfer lithograph in Ives' *Report* (cat. 33f). Ives might have taken his view from his camp (no. 58) in "Giant's Pass," between modern Willow Beach and Crane's Nest Rapids.[11] However, the sketch may have been very crude in the first place, and the print is so generalized and romanticized that the view is probably unrecognizable. The nineteenth-century neo-gothic gloom that pervaded some of Ives' and Möllhausen's written narratives also extends to the print, which has been compared to the works of Gustave Doré.[12]

Ives and his two companions passed beyond the present site of Hoover Dam and reached what they believed to be the mouth of the Virgin River (actually Las Vegas Wash), before turning back on the morning of March 11. Floating back downriver with the current, they arrived at the steamboat camp by nightfall that same day.[13] In the meantime, for Möllhausen and the other men, time had "passed unimaginably slowly," and most of them evidently felt no regret on March 13 when the entire expedition started downriver on the *Explorer* to meet the mule train.[14]

1 Ives, *Report*, part 1, pp. 81-82.
2 Möllhausen, *Reisen*, chapter 17; Ives, *Report*, part 1, p. 82.
3 Ives, *Report*, part 1, pp. 82-83.
4 Möllhausen, *Reisen*, chapter 17.
5 Newberry in Ives, *Report*, part 3, p. 40.
6 Möllhausen, *Reisen*, chapter 17.
7 Compare USGS 1:250,000 map "Kingman" and 7.5' series "Willow Beach" and "Fire Mountain" quadrangles with plats 49 and 50 of Ives' original map and Egloffstein's printed map 1. The site should not be confused with the site marked "Head of Navigation" on the latter, which instead refers to the furthest point Lieutenant Ives reached on his reconnaissance in the skiff.
8 Möllhausen, *Reisen*, chapter 17.
9 Ibid. See also Ives, *Report*, part 1, p. 83.
10 Ives, *Report*, part 1, pp. 85-86.
11 Compare plat 52 of Ives' original map with USGS 7.5' series "Willow Beach" and "Ringbolt Rapids" quadrants.
12 Goetzmann, *Army Exploration in the West*, p. 389; Patricia Trenton and Peter Hassrick, *The Rocky Mountains: A Vision for Artists in the Nineteenth Century* (Norman: University of Oklahoma Press in association with the Buffalo Bill Historical Center, 1983), p. 103.
13 Ives, *Report*, part 1, pp. 86-87.
14 Möllhausen, *Reisen*, chapters 17 and 18.

CAT. No. 34

Balduin Möllhausen
Second Camp After Leaving the Colorado (Railroad Pass)
Watercolor and gouache on paper
8⅜ x 11 in. (21.3 x 27.9 cm)
signed l.r.: "Möllhausen"
l.l. on paper mount: "46"
l.c. on paper mount: "2nd camp after leaving the Colorado"
ACM 1988.1.44

Cat. No. 34

On March 23, Lieutenant Ives and his smaller overland party—including Möllhausen, Dr. Newberry, Egloffstein, Peacock the mule driver, and Lieutenant Tipton's escort of twenty to twenty-four soldiers—set out by land from Beale's Crossing, headed east in search of another route to the Mormon Road. Guided by Iretéba and two or three of his Mohave comrades, they crossed the Black Mountains through a pass traversed earlier by Beale and his camel corps.[1] According to Ives, they formed their first camp in "a snug meadow carpeted with good grass, and fringed on one side by a growth of willows that bordered" a stream, where their "half-starved animals would hardly allow the saddles to be removed in their impatience to enjoy the unaccustomed plenty."[2]

From their camp at Meadow Creek (no. 61) they could see the conical Iretéba's Mountain (cats. 34b-c), where their Mohave guide apparently went ahead to try to locate some Hualapai guides.[3] The expedition members remained at the campsite until the morning of March 26, while their Indian guides visited "a few Apache families who lived deeper in the mountains and who could provide information about the adjacent country."[4] Ives and his men then crossed the arid Sacramento Valley, heading northeast toward the Cerbat Mountains.[5] The lieutenant observed:

The pass by which we were to cross the Cerbat mountains was apparent as soon as we left the Black range, and Ireteba, who had joined us early in the morning, headed directly for it.

The pure atmosphere made it seem close by, and it was disappointing to plod through the hot sand hour after hour, and find it appearing as far off as ever. When the base of the mountains was at last reached, it was found that the ascent was scarcely perceptible. A place more like a cañon than an ordinary mountain pass presented itself, and we penetrated the range for a few miles through the windings of a nearly level avenue. In a pretty ravine, hemmed in by picturesque bluffs, our guide pointed out a good spring of water, with grass enough near by to afford a tolerable camping place.[6]

The camp in this canyon (camp no. 62) forms the subject of Möllhausen's watercolor and the related print (cat. 34a). Two pyramidal tents appear among huge boulders in the center and right foreground, as well as several figures and a campfire. In his diary the artist described the vicinity of the camp as having "scanty bunch grass" growing "sparsely among volcanic debris on the mountain slopes." These "jumbled and crowded" mountains

34A
John James Young after Balduin Möllhausen, *Railroad Pass*, engraving, from Ives, *Report* (1861), part 1, p. 95, Amon Carter Museum Library

looked like plateaus which were separated from one another by canyons and clefts. Huge granite boulders lay scattered in the ravines or towered up out of the ground or projected up from the base of the mountains. But the undisturbed rocks of the mesas towered no more than four to six hundred feet above the valley floor. They consisted of broad layers of red sandstone and conglomerate, covered by a layer of basalt.[7]

Möllhausen also wrote: "By the way, we were not the first whites to have camped there, for we saw unmistakable signs that Lieutenant Beale had been there with his camels."[8]

According to Ives, they named this pass Railroad Pass, following up Lieutenant Whipple's surmise that a

> direct route from the divide to the Colorado would be practicable for a railroad, besides greatly shortening the distance. The observations of the past two days have demonstrated the accuracy of his judgement. From the divide the road can follow the rim of the basin along an unbroken ridge to Railroad Pass, from which place there is a smooth slope to the Colorado.[9]

Today the town of Kingman, Arizona, stands at the northeastern end of the pass, which serves as the route for the A.T.& S.F. Railroad, Interstate 40, and old U.S. Highway 66 (now Arizona State Highway 66).[10]

From Railroad Pass, Iretéba guided the expedition north on March 27 "for ten or fifteen miles along the eastern base of the Cerbat range, to an excellent grazing camp, but where there was only a small spring of sulphurous water."[11] Ives called this area the Cerbat Basin, and Newberry apparently referred to it as the "Yampai Valley"; located in northern Mohave County, it is known today as the Hualapai Valley. After camping there (camp no. 63) for an extra day to water and rest the animals, the expedition trekked across the dry basin in a northeasterly direction, following Beale's trail around the northern end of the Peacock Mountains (named for their mule driver), more or less along modern Arizona Route 66 and the A.T. & S.F. Railroad.[12]

Between Crozier and Truxton, according to Ives: "Mr. Peacock, riding in advance, discovered a large spring of clear, sweet water in a ravine near the road. There were no signs of the place having been used as a camp, and even Iretaba did not appear to have

34B
John James Young after Balduin Möllhausen, F. W. von Egloffstein, or John Strong Newberry, *Ireteba's Mountain*, engraving, from Ives, *Report* (1861), part 1, p. 94, Amon Carter Museum Library

known previously of its existence."[13] The men spent the nights of March 30 and 31 (camp 65) at this spring, apparently the one now called Truxton Spring. During the second day, Egloffstein went hunting with Iretéba while Möllhausen hunted with Hamotamaque, one of Iretéba's Mohave companions. The expedition also first encountered Hualapais Indians at this camp.[14]

The original sketch for the print titled *Colorado Plateaus from near Peacock's Spring* (cat. 34d) could date from Möllhausen's hunting trip, when he climbed "up the heights to the south," or from the next day, when, a mile or two toward the future site of Truxton, Ives reported:

> we issued from the hills and entered a region totally different from any that had been seen during the expedition. A broad table-land, unbroken by the volcanic hills that had overspread the country since leaving Fort Yuma, extended before us, rising in a gradual swell towards the north. The road became hard and smooth, and the plain was covered with excellent grass. Herds of antelope and deer were seen bounding over the

34c
Ireteba's Mountain in March 1990

they would continue to respond to the awesome beauty and the marvelous formations they encountered.

1 Ives, *Report*, vol. 1, p. 91, and Möllhausen, *Reisen*, chapter 21. Lieutenant Ives mistakenly believed that Sitgreaves had also crossed the mountains at this point, although according to Wallace, "Sitgreaves' Expedition," p. 353, and Barnes, *Arizona Place Names*, p. 409, Sitgreaves used "Union Pass" along State Highway 68; compare Egloffstein's printed map 2 with USGS 1:250,000 map "Kingman." Long before these explorers, however, in early June 1776, Francisco Garcés may have crossed the Black Mountains by the same pass Beale and Ives later used. See Coues, ed., *On the Trail of . . . Garcés*, vol. 2, p. 315.
2 Ives, *Report*, part 1, pp. 93-94.
3 Although the name Ireteba's Mountain does not appear on modern maps, the landmark can still be seen. Ives (*Report*, part 1, pp. 94-95) described it as "a conical hill four or five hundred feet high, surmounted by a cylindrical tower. It is a conspicuous feature among the other summits, and would be a good landmark to guide the traveller from the east to the pass, and to an excellent camping place at its mouth." Also see Newberry in Ives, *Report*, part 3, p. 49.
4 Möllhausen, *Reisen*, chapter 21.

slopes. Groves of cedar occurred, and with every mile became more frequent and of larger size. At the end of ten miles the ridge of the swell was attained, and a splendid panorama burst suddenly into view. In the foreground were the low table-hills, intersected by numberless ravines; beyond these a lofty line of bluffs marked the edge of an immense cañon; a wide gap was directly ahead, and through it were beheld, to the extreme limit of the vision, vast plateaus, towering one above the other thousands of feet in the air, the long horizontal bands broken at intervals by wide and profound abysses, and extending a hundred miles to the north, till the deep azure blue faded into a light cerulean tint that blended with the dome of the heavens. The famous "Big cañon" was before us; and for a long time we paused in wondering delight, surveying this stupendous formation through which the Colorado and its tributaries break their way.[15]

The print titled *Colorado Plateaus from near Peacock's Spring* is probably the first view in Ives' *Report* to show the distant canyonlands.

As they continued northeast along Beale's road, "numerous canyons began to cut through the plateau in a northwesterly direction," Möllhausen noted, and "the rocky canyon walls sank deeper and deeper until in the distance they blurred together in a violet haze."[16] The explorers were most interested in finding a route through or around the canyonlands, but in the days ahead

34d
John James Young after Balduin Möllhausen or F. W. von Egloffstein,
Colorado Plateaus from near Peacock's Spring, engraving,
from Ives, *Report* (1861), part 1, p. 98, Amon Carter Museum Library

5　According to Barnes, *Arizona Place Names*, p. 373, the Sacramento Valley and Wash in Mohave County was "named for Sacramento, Calif., by miners who came to Arizona with California regiments." Ives and his men never used this name. Newberry (in Ives, *Report*, part 3, p. 90) called it "Long Valley," "an interval of nearly level land some fifteen miles wide, lying between the Black mountains and the next succeeding range on the east. It extends north nearly to the Colorado, . . . parallel to . . . the Mojave valley, but . . . much higher, (2,000 feet,) . . . This plain forms the first step in the ascent to the table-lands."

6　Ives, *Report*, part 1, p. 95. The mountains that Ives and his men referred to collectively as the Cerbat Mountains are now known as the Cerbat Mountains from Kingman northward and as the Hualapais Mountains southeast of Kingman.

7　Möllhausen, *Reisen*, chapter 21.

8　Ibid. Coues, ed., *On the Trail of . . . Garcés*, vol. 2, pp. 317-319, speculated from Garcés' vague diary entry for June 7, 1776, that the latter also may have been in the Railroad Pass vicinity, where he encountered some "Jaguallapais" (Hualapais) Indians.

9　Ives, *Report*, part 1, pp. 95-96.

10　Compare Egloffstein's map with USGS 1:250,000 map "Kingman."

11　Ives, *Report*, part 1, p. 96.

12　Compare Egloffstein's map with USGS 1:250,000 map "Williams." Spanish priest and explorer Francisco Garcés had probably traversed this route in June 1776. See Coues, ed., *On the Trail of . . . Garcés*, vol. 2, p. 322. In addition, the Sitgreaves expedition had passed through the vicinity in late October and early November of 1851. See Wallace, "The Sitgreaves Expedition," pp. 349-351, map betw. pp. 342 and 343.

13　Ives, *Report*, part 1, p. 97. In addition to Peacock Spring and Peacock Mountains, there is a Peacock Peak, according to Barnes, *Arizona Place Names*, p. 323. However, as Barnes notes, Coues in *On the Trail of . . . Garcés*, vol. 2, pp. 322-323, believed that Peacock Spring and Truxton Spring were one and the same, and that USGS and GLO maps had placed modern Peacock Springs "twelve miles off" from the location that Ives' map had described. Be that as it may, Ives camped at Truxton Spring, named by Lieutenant Beale for his maternal grandfather, Commodore Thomas Truxtun. Sitgreaves also camped here from October 30 to November 2, 1851. The history of this important water is summarized in Granger, *Arizona's Names*, pp. 627-628; also see Wallace, "Sitgreaves Expedition," pp. 350-351.

14　Möllhausen, *Reisen*, chapter 22.

15　Ives, *Report*, part 1, pp. 98-99. Ives makes no mention here of passing the Music Mountains, which Möllhausen notes they "christened . . . because of their appearance." He described them as "mesa-like remnants of a high plateau" which "rose up about one thousand feet above their bases" and "consisted of numerous horizontal layers of solid rocks and more tractable masses of earth." He wrote: "Cedar trees dotted the slopes on the layers of soil, while the rock layers were bare. Since these parallel lines were repeated at peculiarly regular intervals, the rugged declivities bore a striking similarity to lines of musical notes on paper." Möllhausen, *Report*, chapter 22.

16　Möllhausen, *Report*, chapter 22.

Cat. No. 35

Balduin Möllhausen
Entrance of Wallpay Cañon [Peach Springs Canyon]
Watercolor and gouache on paper
8¼ x 11 in. (21.0 x 27.9 cm)
signed l.r.: "Möllhausen"
l.l. on paper mount: "8"
l.c. on paper mount: "Entrance of Wallpay cañon"
verso, center on paper mount: "Entrance to Wallpay Cañon"
ACM 1988.1.8

On April 1, the expedition left Beale's Road at the site of the modern town of Peach Springs, Arizona, and headed north. Hualapai guides led them through intricate and narrow ravines in a great descent along a blind and circuitous trail. Ives reported:

Cat. No. 35

> A few miles of difficult travelling brought us into a narrow valley flanked by steep and high slopes; a sparkling stream crossed its centre, and a gurgling in some tall grasses near by announced the presence of a spring. The water was delicious. The grass in the neighborhood was sparse, but of good quality.[1]

According to Möllhausen, "around evening" they reached a "clear but tepid spring" from which "water rushed a few hundred feet . . . into the canyon and then sank into the sand." He noted cottonwood trees, scrub oak, and a lone peach tree in the vicinity. "We speculated about how the peach stone might have found its way to this secluded region," he wrote. "We assumed it had come from the Rio Grande and was planted by Apache Indians by chance."[2]

The expedition camped (no. 66) near this spring (designated as "New Creek" on Egloffstein's map and today known as Peach Springs, 3½ miles from town on the Hualapai Indian Reservation).

Möllhausen described his first view of the entrance to "Wallpay Cañon," now known as Peach Springs Canyon[3]:

Ahead of us, about a mile away, the canyon was hemmed in by high vertical rock walls which showed . . . regular horizontal layers and strata. . . . The actual mouth of this mysterious canyon, where our road would lead the next day, was concealed by some hills, so I climbed a nearby bluff to make a sketch of this interesting place from that vantage point. A maze of regular and irregular lines—of which the former predominated—were concentrated together in a peculiar, but beautiful scene. The rugged plateaus pushed past one another like colossal ramparts with horizontal walls. Their slopes glistened in multi-colored lines which were virtually horizontal, and darker shadows revealed places that went down deep into the bowels of the earth. Cedar bushes adorned the adjacent rolling hills which were stacked one upon another. Barren rocks rose up behind them in the most magnificent formations, and because of variations in distance, they were shrouded in the most delicate gradation of colors which blended together at a distance. Silent tranquility reigned in this desolate, yet beautiful wilderness. But the lifeless rock, the green cedars and the germinating blades of grass spoke to those who were receptive in an easily understandable way: "Nature is sublime in all her forms!"[4]

He made his original sketch for this watercolor, *Entrance to Wallpay Cañon*, on a hilltop a few yards away from a spring where the expedition camped on the night of April 1 (camp no. 66). The watercolor, which looks north or northeast, depicts the landscape at sunset, when deep shadows, contrasts, and coloration add to its beauty and interest. The camp and spring were located beyond the picture, in the valley at the right. Möllhausen further recorded that after making the sketch he "took the nearest path down a steep slope" to the spring, while "the rocky surface gave way under . . . [his] feet." Since he was careful to note that "pieces of sandstone, marble, and soft shale mixed with weathered fragments of limestone" rolled down ahead of him, he may have had Dr. Newberry with him.[5] The two figures in the watercolor's foreground may be Möllhausen and the geologist.

While they were in camp that night, six or seven Hualapais came to visit and converse as they all sat by the campfires. The next morning the expedition left the valley, following the course of the creek down a ravine, both of which appear in shadow in the center of the picture. Although the creek soon disappeared

around the small, unusual buttes in partial shadow just to the left of center, the ravine continued around to the left and led them into the canyon, which continues out of the picture at far left.[6]

A reproduction of Möllhausen's watercolor did not appear in Ives' *Report*. The Berlin Geographical Society may at one time have had another version of this view, since the 1904 list of Möllhausen's watercolors includes number 40: *"Nach Verlassen des Stromes auf der Hochebene der Eingang der Schluchten nach dem Diamond-Thal hinunter* ("The Entrance to the Canyons to the Diamond Valley on the High Plateau After Leaving the River.")

An engraved view in Ives' *Report*, titled *Side Cañons of Diamond River* (cat. 35a), shows a recognizable peak from the Hualapai Indian Reservation in northern Arizona, and a recent photograph of the formation taken from the floor of Peach Springs Canyon (cat. 35b) suggests that the artist made the original sketch from an elevated position along the canyon's walls.[7] (The engraver, rather than the artist, probably sharpened the inclines of the peak and

35A
John James Young after Balduin Möllhausen or F. W. von Egloffstein,
Side Cañons of Diamond River, engraving, from Ives, *Report* (1861),
part 1, p. 99, Amon Carter Museum Library

35ʙ
View from the floor of Peach Springs Canyon, March 1990

that in the Black cañon, excepting that the rapid descent, the increasing magnitude of the collossal [*sic*] piles that blocked the end of the vista, and the corresponding depth and gloom of the gaping chasms into which we were plunging, imparted an unearthly character to a way that might have resembled the portals of the infernal regions. . . . At short distances other avenues of equally magnificent proportions came in from one side or the other; and no trail being left on the rocky pathway, the idea suggested itself that were the guides to desert us our experience might further resemble that of the dwellers in the unblest abodes—in the difficulty of getting out.[10]

Compounding that sense of unease was the proximity of the Hualapais, whose presence was marked by "[h]arsh screams issuing from aerial recesses in the cañon sides and apparitions of goblin-like figures perched in the rifts and hollows of the impending cliffs."[11]

In a side canyon, the white men passed a camp of huts "of the rudest construction, visible here and there in some sheltered niche or beneath a projecting rock," according to Ives; Möllhausen estimated that the camp contained thirty Indians.[12] At some point, the artist made both a field sketch and a watercolor depicting these "mysterious" people (cat. 35c and p. 131), and a lithograph titled *Hualpais* appeared in Ives' *Report*. The whites' romantic European sensibilities probably predisposed them to see the Hualapais and the sublime scenery through a gothic gloom. In their turn, the Hualapais, by their shyness, apparently saw the expedition's intrusion into their domain with wariness and suspicion. When the visitors encountered a "hideous old squaw, staggering under a bundle of fuel," Ives noted:

Our party being, in all probability, the first company of whites that had ever been seen by them, we had anticipated producing a great effect, and were a little chagrined when the old woman, and two or three others of both sexes that were met, went by without taking the slightest notice of us. If pack-trains had been in the habit of passing twenty times a day they could not have manifested a more complete indifference.[13]

Ives wrote of his apprehensions and those of the men before they reached Diamond Creek: "Seventeen miles of this strange travel had now been accomplished. The road was becoming more

compressed the framing canyon walls to fit the rectangular format of the plate.) The expedition first viewed this formation on April 2, after travelling the length of Peach Springs Canyon. Dr. Newberry described the topography with the obvious enthusiasm of a geologist in his element,[8] but the rest of the men experienced the canyon with a less scientific interest. Möllhausen wrote of "temples of wonderful architecture, long colonnades and colossal, but delicate pyramids," and "vast vaults, arched windows and gates."[9] According to Ives:

The bottom was rocky and irregular, and there were some jump-offs over which it was hard to make the pack animals pass. The vegetation began to disappear, leaving only a few stunted cedars projecting from the sides of the rugged cliffs. The place grew wilder and grander. The sides of the tortuous cañon became loftier, and before long we were hemmed in by walls two thousand feet high. The scenery much resembled difficult, and we looked ahead distrustfully into the dark and apparently interminable windings, and wondered where we were to find a camping place."[14] The Hualapais' customary route, he noted, "follows the cañon of Diamond creek, but this they pronounced impracticable for mules, and said that we must retrace our course for several miles in order to strike a more circuitous but easier trail, that ascended one of the branch cañons."[15]

On April 4, after they had spent two nights in a camp (camp 67) just behind an outcrop of canyon wall (see the left center of cats. 35a and 35b), the expedition members retraced their steps halfway back up Peach Springs Canyon.[16]

---

1 Ives, *Report*, part 1, p. 99.

2 Möllhausen, *Reisen*, chapter 22. Coues, *On the Trail of . . . Garcés*, vol. 2, pp. 327-328, n.30, maintains that Peach Springs is what Garcés called "Pozos de San Basilio"—St. Basil's Wells. But careful translation of Garcés' account indicates no digression from the main course of Truxton Wash, which would have been necessary to reach the water in Möllhausen's Wallpay Canon. Nor are these springs a *pozo*—a well or pothole. Beale apparently was the first non-Indian to describe the springs in print; he called them Hemphell's Springs after visiting in the summer of 1858. This spot became a favored camping place on the Beale Road, but Möllhausen is the only visitor to mention the peach tree. It is possible that Ives' party were the first white men to have camped there. See Granger, *Arizona's Names*, p. 469.

3 Barnes, *Arizona Place Names*, pp. 321-322. Compare Egloffstein's map 2 with USGS 1:250,000 map "Williams" and USGS 7.5' series map "Peach Springs Quadrangle" (1967).

4 Möllhausen, *Reisen*, chapter 22.

5 Ibid. In Ives, *Report*, part 3, p. 54, Dr. Newberry described this area as a plain that drained

> toward the north, and in that direction the arroyos become deeper and expose thick beds of sand, gravel, and boulders; the greater part of the pebbles being of a hard bluish-gray limestone. Ultimately the beds of the draining streams were found to have reached the rocky floor beneath. This they soon begin to erode, forming cañons which lead toward the Colorado. These cañons gradually increase in number and depth till twenty miles north of the point where the denudation commences, the plateau is intersected by a labyrinth of chasms from 2,000 to 3,000 feet deep, of which the nearly perpendicular walls forbid the passage of man or any other wingless animal.

6 Möllhausen, *Reisen*, chapter 22. Also see Ives, *Report*, part 1, pp. 99-100.

7 Compare Egloffstein's map with USGS 1:250,000 map "Williams" and USGS 7.5 min. series maps "Peach Springs Canyon Quadrangle" and "Peach Springs NE Quadrangle" and "Diamond Peak Quadrangle." Pioneer photographer George Benjamin Wittick (1845-1903) and his son Tom travelled Ives' exact route from

35c
Balduin Möllhausen, *Two Warriors (Hualapais) [Zwei Krieger]*, graphite on paper, from Sketchbook 3, Möllhausen Family, Bleicherode, Germany, photograph courtesy Dr. Friedrich Schegk, Munich

the Peach Springs station of the A.T.& S.F. Railroad to the mouth of Diamond Creek in 1883, and photographed the exact peak (Sentinel Peak?) at the junction of Peach Springs and Diamond Canyons. It is reproduced in Terence Murphy and Tom Wittick, "An 1883 Expedition to the Grand Canyon: Pioneer Photographer Ben Wittick Views the Marvels of the Colorado," *American West* 10 (March 1973), p. 43.

8   In Ives, *Report*, part 3, p. 54, he observed:

> Near Camp 66 we descended to the bottom of one of the arroyos . . . where it has a depth of more than 500 feet. A mile north of this point its sides converge and become perpendicular, forming a cañon of more magnificent proportions than any we had seen, and in geological interest far surpassing anything I had dared to hope for.
>
> This seemed to be our only avenue of approach to the Colorado, and we followed it for fifteen miles, to its junction with the still grander cañon of that stream.

9   Möllhausen, *Reisen*, chapter 22.

10 Ives, *Report*, part 1, pp. 99-100.

11 Ives, *Report*, part 1, p. 100.

12 Ibid.; Möllhausen, *Reisen*, chapter 22.

13 Ives, *Report*, part 1, p. 100.

14 Ibid.

15 Ibid.; see Egloffstein's map, camp 67.

16 Möllhausen, *Reisen*, chapter 23.

CAT. NO. 36

Balduin Möllhausen
Camp on Diamond Creek
Watercolor and gouache on paper
8⅝ x 11⅛ in. (22.0 x 28.3 cm)
signed l.r.: "Möllhausen"
l.l. on paper mount: "45"
l.c. on paper mount: "Camp on Diamond Cr."
ACM 1988.1.43

On April 4, as they continued through Peach Springs Canyon, Lieutenant Ives wrote:

Cat. No. 36

> At last we struck a wide branch cañon coming in from the south, and saw with joyful surprise a beautiful and brilliantly clear stream of water gushing over a pebbly bed in the centre, and shooting from between the rocks in sparkling jets and miniature cascades. On either side was an oasis of verdure—young willows and a thick patch of grass. Camp was speedily formed, and men and mules have had a welcome rest after their fatiguing journey.
>
> A hundred yards below camp the cañon takes a turn; but as it was becoming very dark, all further examinations were postponed till to-morrow. In the course of the evening Ireteba came into my tent, and I asked him how far we had still to travel before reaching the great river. To my surprise he informed me that the mouth of the creek is only a few yards below the turn, and that we are now camped just on the verge of the Big Cañon of the Colorado."[1]

According to Möllhausen, they named the creek Diamond Creek, probably because its water was "pure and clear as a diamond." Comparing the "charming little valley" to "a precious stone," he described how, at this oasis: "We forgot both the troubles of the day and the Colorado, and feeling almost euphoric, we hurried to the base of an overhanging rock wall to erect our tents on the soft grass. Everyone was happy. The men joked as they did their camp duties and the animals grunted with pleasure as they rolled in the lush grass."[2]

Möllhausen's watercolor *Camp on Diamond Creek* shows the bubbling waters of the creek and the colossal peak opposite its mouth on the Colorado River. A recent photograph of the location (cat. 36b) demonstrates the essential topographical accuracy of Möllhausen's view; the creek, although obscured by vegetation, still flows in the lefthand portion of the photograph.[3] As in all parts of the Grand Canyon, the time of day and atmospheric effects strongly influence the shading and tonalities of the rocky surfaces.

Möllhausen recorded when and where he made his original sketch:

Immediately after our arrival I climbed up a projecting wall, and there I gained a view of the little glen and its enclosing mountains, and made a sketch of the beautiful scene. The butte was over two thousand feet high, and appeared to hem in Diamond Creek to the northwest. I had a great deal of difficulty sketching the various lines on the mountain. I could not see its base because a projecting point obstructed my view. It resembled a mighty unfinished edifice surrounded by battlements and towers. I recognized uniform strata which, deposited in layers, extended like artificial masonry across the entire face of the huge rock and all the way to its summit. The atmosphere and occasional rains fashioned it into such singular formations. Similar mountains appeared in the distance on all sides. If one followed the different layers of strata, it was easy to see that the mountains, which were now widely separated, were once closely connected, forming a solid plateau.

The steep walls offered a strange play of colors. The first eight hundred feet were predominantly dark-brown and blue-black, while the upper strata glowed in a most beautiful rose, yellow, blue or green, depending upon how the setting sun picturesquely illuminated these formations, which had deposited one upon the other across the ages. The unusual clarity of the air made distant objects appear much closer than they actually were, leading all of us to believe that the beautiful butte which I have just described was separated from us only by a rocky projection. However, when I returned to camp I encountered Iretéba, who assured me that the Colorado flowed between us and that mountain.[4]

36A
John James Young after Balduin Möllhausen, *Near Head of Diamond Creek*, toned lithograph, from Ives, *Report* (1861), part 3, opp. p. 54, Amon Carter Museum Library

36B
The view along Diamond Creek, March 1990

36c
After Balduin Möllhausen, *Der Diamant-Bach (Diamond Creek)*, toned, electrotype engraving (hand colored) or transfer lithograph, from Möllhausen, *Reisen* (Leipzig: Otto Purfurst, 1861), vol. 2, opp. p. 48

Even Dr. Newberry resorted to architectural allusions to describe the rocky peaks or buttes in the vicinity:

> Near the mouth of Diamond river, by the intersections of the numerous cañons which cut the plateau, portions of it have been left in a series of pinnacles and pyramids, frequently standing entirely isolated, forming some of the most striking and remarkable objects seen on our expedition. Many of these buttes exhibit a singular resemblance to the spires and pyramids which form the architectural ornaments of the cities of civilized nations, except that the scale of magnitude of all these imitative forms is such as to render the grandest monuments of human art insignificant in comparison with them.[5]

Möllhausen's view became the basis for two separate prints: a lithograph for Ives' *Report* (cat. 36a) and a lithograph for the second volume of *Reisen* (cat. 36c). In redrawing Möllhausen's sketch for Ives, John James Young attempted to show how the colossal peak in the center of the picture formed part of the Grand

Canyon's northwest wall. Young leveled the strata in the distant formation, making it seem that the original view was made from an even higher elevation, and exaggerated the height of the granite pinnacles in the middle distance. The German lithograph more closely approximates the topography of Möllhausen's sketch, and tents have been added in the center foreground, perhaps with the approval of the artist.

Using the camp at Diamond Creek as a base, the men explored in several different directions and sketched many nearby wonders (see cat. nos. 37 and 38).[6] At least one of Möllhausen's related images was lost or destroyed in World War II; in 1904 the artist gave the Berlin Geographical Society a watercolor of the "rocky stream bed of Diamond Creek" (1904 list number 41: *Felsenbett des Diamond Creek*).

1   Ives, *Report*, part 1, p. 100.
2   Möllhausen, *Reisen*, chapter 23.
3   Compare Egloffstein's map 2 (camp no. 67) and USGS 1:250,000 map "Williams." Terrence Murphy and Tom Wittick, in "An 1883 Expedition to the Grand Canyon," p. 42, reproduce Ben Wittick's 1883 photograph of the same subject, taken from almost the same location as Möllhausen's view.
4   Möllhausen, *Reisen*, chapter 23.
5   Newberry in Ives, *Report*, part 3, p. 55; that page also contains Newberry's detailed geological cross-section of the canyon walls at camp 67.
6   On the afternoon of April 3, Möllhausen and Newberry may have walked to the geological formations shown in *Granite Pinnacles, Cañon of Diamond River*, which Dr. Newberry probably sketched a few miles up Diamond Creek Canyon. See Ives, *Report*, part 3, p. 57.

CAT. NO. 37

Balduin Möllhausen
Mouth of Diamond Creek on Colorado River, View from the North
Watercolor and gouache on paper
8⅝ x 11¼ in. (22.0 x 28.6 cm)
signed l.r.: "Möllhausen"
l.l. on paper mount: "33"
l.c. on paper mount: "Mouth of Diamond Creek on Colorado
river./ view from the north"
verso, center on paper mount: "Mouth of Diamond Creek &
Colorado River/ View from th[*sic*] North"
ACM 1988.1.31

CAT. NO. 38

Balduin Möllhausen
Mouth of Diamond Creek, Colorado River, View from the South
Watercolor and gouache on paper
8⅜ x 11⅛ in. (21.3 x 28.3 cm)
signed l.r.: "Möllhausen"
l.c. on paper mount: "Mouth of Diamond Creek.
Colorado river view from the south"
verso, center on paper mount:
"Mouth of Diamond Creek/Colorado River viewed from the South"
ACM 1988.1.36

Möllhausen recorded that near evening on April 2, he and several of the men left their new camp on Diamond Creek to explore the area nearby. As they followed the stream toward the butte the artist had sketched (cat. no. 36), the canyon of Diamond Creek "gradually opened into a broad sandy valley, and the base of the renowned butte appeared to rise up there from the sandy floor." Although it seemed unlikely that the Colorado could be there, they then heard "a muffled, mighty, wild, thundering like the repeated stamping of countless ponderous hoofs." The men rushed onward through underbrush and the sandy flats until they saw the river's "foaming surface," plunging "with irresistible force over debris which it had broken loose." Möllhausen wrote that the view left them "speechless" in the "solemn tranquility" of "the

majestic formations." However, the sun had set, and the increasing darkness allowed them to revel "for only a moment in the view of this grandiose scene" before they had to head back to camp.[1]

The next morning Möllhausen and Newberry returned to the river, while Lieutenant Ives walked there for the first time. The commander reported that

Cat. No. 37

the scene was sufficiently grand to well repay for the labor of the descent. The cañon was similar in character to others that have been mentioned, but on a larger scale, and thus far unrivalled in grandeur. Mr. Möllhausen has

Cat. No. 38

taken a sketch, which gives a better idea of it than any description. The course of the river could be traced for only a few hundred yards, above or below, but what had been seen from the table-land showed that we were at the apex of a great southern bend. The walls, on either side, rose directly out of the water. The river was about fifty yards wide. The channel was studded with rocks, and the torrent rushed through like a mill-race.[2]

Möllhausen, in fact, made two sketches, one from a ledge to the north of the confluence and another from rocky debris to the south. While Dr. Newberry "diligently hammered about among the rocks," Möllhausen "looked for a suitable spot for sketching, where I could obtain a full, beautiful view of the picturesque rocky portal through which the foaming Colorado plunged."[3] His first sketch served as a model for the watercolor *Mouth of Diamond Creek on Colorado River, View from the North* (cat. no. 37). After he completed this sketch, Möllhausen

set out toward the southern end of Diamond Creek delta to make a sketch of the northern gorge as well; and there I sat in the cool shadows of an overhanging granite wall. The violent surf foamed nearby, while the sun's rays shone brightly through the wide opening of Diamond Creek on the rolling water and on the rugged rocks on the opposite bank. A few soldiers and Mexicans crouched on the bank fishing, and a lone heron passed by me slowly beating its wings. In the loud roar of falling water the bird's hoarse cry faded away along with the sound of the men's laughing and joking, and a characteristic impression of undisturbed solitude settled over this terribly beautiful wilderness.

It was long past noon when I took my rifle and sketchbook and got ready to return to camp. I looked one last time at the Colorado's foaming waves and then wandered slowly up Diamond Creek."[4]

The second watercolor, *Mouth of Diamond Creek, Colorado River, View from the South* (cat. no. 38), complements the first one; in fact, each shows the site where the artist made the original sketch for the other view. Recent photographs taken in the vicinity (cats. 37a and 38a) show that the watercolors and the narrative descriptions reproduce the topography with reasonable accuracy.[5] The view taken from the north shows the canyon of Diamond Creek, with its sandy floor and vegetation, entering at left, just beyond the foreground precipice. This same precipice is also visible in the view taken from the south, this time just right of the narrow center passage. This latter view shows Diamond Creek Canyon entering from the right, just beyond the dark foreground rocks that correspond to the rocky lower canyon walls just left of center in the other view. Both watercolors use these dark, rocky foreground projections and figures as répoussoir elements to frame the images and provide a sense of scale, while the rapids are effectively rendered through the use of opaque white.

Möllhausen listed a related watercolor titled *Mündung des Diamond Creek in den Colorado* ("Mouth of Diamond Creek on the Colorado") as number 40 on the list of watercolors he sent to the Berlin Geographical Society in 1904. He also noted that it hung framed in his study. It is impossible to determine from its title whether this view, which was lost or destroyed in World War II, was taken from the north or south, or from a different location entirely. A view from the south served as a model for the print in *Reisen* titled *Der Rio Colorado, nahe der Mündung des Diamant-Baches* ("The Colorado River near the Mouth of Diamond Creek") (cat. 38b); its topography is quite similar to that in the Amon Carter

37A
Mouth of Diamond Creek, March 1990

38A
Mouth of Diamond Creek, March 1990

38ʙ

After Balduin Möllhausen, *Der Rio Colorado, nahe der Mündung des Diamant-Baches*, toned electrotype engraving (or transfer lithograph?), from Möllhausen, *Reisen* (Leipzig: Otto Purfürst, 1861), vol. 2, opp. p. 54

watercolor (cat. no. 38), except that the printmaker added a campfire in the right foreground—apparently for pictorial interest, since it seems to bear no relation to fact.

Egloffstein evidently made a sketch (now lost) of the Grand Canyon at the mouth of Diamond Creek, and J. J. Young redrew it as the print titled *Big Cañon at Mouth of Diamond River* (cat. 38c) in Ives' *Report*. This print distorts the actual topography, exaggerating the steepness of the canyon walls in a manner reminiscent of the sublime engravings of Gustave Doré.[6]

38ᴄ

John James Young after F. W. von Egloffstein, *Big Cañon at Mouth of Diamond River*, transfer lithograph (from an engraving?), from Ives, *Report* (1861), opp. p. 100, Amon Carter Museum Library

1 Möllhausen, *Reisen*, chapter 23.
2 Ives, *Report*, part 1, p. 100.
3 Möllhausen, *Reisen*, chapter 23.
4 Ibid.
5 An 1883 photograph by Ben Wittick, reproduced in Terrence Murphy and Tom Wittick, "An 1883 Expedition to the Grand Canyon," p. 47, depicts almost the same view as that taken from the north by Möllhausen.

6 Goetzmann mentions Dante and Doré in connection with Egloffstein's prints and Ives' description in *Army Exploration*, p. 389; Goetzmann and Goetzmann, *West of the Imagination*, p. 111. Egloffstein is also credited with the panoramic lithograph titled *Big Cañon near Diamond River* (Ives, *Report* [1861], part 1, between pp. 102 and 103). As is usual with his outline panoramas, this print does not portray recognizable topographical features but rather the general character of the country.

CAT. NO. 39

Balduin Möllhausen
Ireteba's Turn
Watercolor and gouache on paper
7¾ x 11 in. (19.7 x 27.9 cm)
signed l.r.: "Möllhausen"
l.l. on paper mount: "20"
l.c. on paper mount: "Iretebas turn"
ACM 1988.1.20

Cat. No. 39

According to the Hualapais who visited Ives and his men at their camp on Diamond Creek, Peach Springs Canyon offered the only practical route for the mules to reach the plateau again. Therefore, on April 4 the expedition went back up Peach Springs Canyon for about six miles, then turned southeast into a branch canyon today known as Hell's Canyon. Although the Hualapai guides apparently attempted to slip away from the expedition, they all eventually reached a camping place with a spring, near the upper entrance to Hell's Canyon.[1]

The watercolor *Ireteba's Turn* may depict the view from this camp (no. 68), probably near present-day Mettuck Spring (which Egloffstein and Ives called "Hualpais Spring") on the Hualapai Indian Reservation.[2] Somewhere in this vicinity, Egloffstein also must have found a place to sketch his panorama titled *Big Cañon from near Hualpais Spring* (cat. 39a). According to Möllhausen, the camp was about 2,500 feet above their previous camp on Diamond Creek.[3]

Here Iretéba, fearing that he and his two Mohave companions should not attempt to travel further into Hualapai territory, regretfully told Ives that it was time for them to return to their tribe back on the lower Colorado.[4] Möllhausen witnessed the presentation of gifts to the Mohave guides that evening in front of the lieutenant's tent, and he wrote that he was "very happy to see that Lieutenant Ives had made such good use of government property by rewarding the Indians so generously." The gifts included two mules, red woolen blankets, white cotton manta, colorful scarves,

39A
F.W. von Egloffstein, *Big Cañon From Near Hualpais Spring*, lithograph, from Ives, *Report* (1861), part 1, between pp. 92 and 93, Amon Carter Museum Library

white glass beads, tobacco, knives, little mirrors, kitchen utensils, food, and many other things. The individual members of the expedition also gave them gifts, with Iretéba receiving the most.[5]

The next morning the Mohaves packed their gifts on their mules, and Iretéba confidentially warned Ives to beware of the treacherous Hualapais. He indicated that the Mohaves "would certainly be watched during their return; and if not vigilant, would lose both their presents and their lives, and that they were going to travel, for two days, without rest or sleep."[6] After shaking hands and saying farewell to each expedition member, the Mohaves departed in a southerly direction. Not long afterward the expedition headed north and east.[7]

On April 5, ascending "a bluff nearly a thousand feet" above their previous campsite (no. 68), Lieutenant Ives and his men and

mules reached the top of the Coconino Plateau (see map, p. 49).
In his entry from "Camp 69, Cedar Forest, April 5," the lieutenant
wrote: "Since attaining the summit the road has been good, and
has traversed a slightly undulating park, covered with luxuriant
grass, and interspersed with cedar groves, where deer, antelope,
and hare have been startled by the approach of the train from their
shady coverts."[8]

Möllhausen wrote that they planned

to reach the Colorado again at its confluence with the Little
Colorado so that we could determine that point astronomically.
Since we knew from several reliable sources that we were at a
latitude where the Colorado makes a sharp bend to the east, we
were in complete agreement with the direction our guides
chose. The plateau, which according to barometric observation rose up over 4,000 feet above sea level, appeared as a wide
rolling plain dotted with cedars which grew abundantly in the

39B
Balduin Möllhausen,
*Cedar or Juniper, April 5th (probably
Juniperus tetragona or Utah juniper)
[Gespaltener Baum]*, graphite
on paper, from Sketchbook 3,
Möllhausen Family,
Bleicherode, Germany,
photograph courtesy of
Dr. Friedrich Schegk, Munich

depressions. We traveled ten miles today and made camp on the edge of a sparse stand of trees, where we had the advantage of a little grass and an abundance of wood.[9]

He footnoted the word "cedars" in this passage in *Reisen*, identifying them as "Juniperus tetragona." One of these is apparently the subject of his finely-rendered sketch (cat. 39b), which the artist or his lithographer used as a model for the cedar at left in the print in *Reisen* titled *Vegetation des Hoch-Plateaus* (cat. 39c).

1 Ives, *Report*, pp. 101-102; Möllhausen, *Reisen*, chapter 23.
2 Compare Egloffstein's printed map 2 with USGS 7.5' series map "Peach Springs Canyon Quadrangle" (1967) and USGS 1:250,000 map "Williams."
3 Möllhausen, *Reisen*, chapter 23; also see Ives, *Report*, part 1, p. 102.
4 Ives, *Report*, part 1, p. 102.
5 Möllhausen, *Reisen*, chapter 23.
6 Ives, *Report*, part 1, p. 102.
7 Möllhausen, *Reisen*, chapter 23.
8 Ives, *Report*, part 1, p. 102.
9 Möllhausen, *Reisen*, chapter 23. In a later and especially descriptive passage, he remarked that "the distant sparse cedar forests looked like clusters of black spots similar to a scattered herd of grazing cattle."

39c
After Balduin Möllhausen, *Vegetation des Hoch-Plateaus*, toned lithograph, from Möllhausen, *Reisen* (Leipzig: Otto Purfurst, 1861), vol. 2, opp. p. 222

Cat. No. 40

Balduin Möllhausen
Character of the High Table Lands (Camp—Colorado Plateau)
Watercolor and gouache on brown-toned paper
8½ x 11¾ in. (21.6 x 29.9 cm)
signed l.r.: "Möllhausen"
l.l. on paper mount: "48"
l.c. on paper mount: "Character of the high table lands"
ACM 1988.1.46

Cat. No. 40

Möllhausen's fine watercolor titled *Character of the High Table Lands (Camp—Colorado Plateau)*, the original field sketch that inspired it *(Landscape on the High Plateau, April 10th)*, and his narrative description in *Reisen* demonstrate the interrelation of his works. The field sketch (cat. 40a) recorded the essential landscape components; for the finished watercolor, the artist added figures, campfires, a tent, and some dead trees, incorporating these additional details from other sketches (now lost) or from his written notes. Möllhausen's skillful use of diagonal strokes to indicate foliage in the graphite sketch finds even greater expression in the finished watercolor, where the artist employed a combination of opaque white and drybrush techniques to portray snow and transparent gray washes for smoke. The whitish yellow-orange of the campfires is reflected in the men's faces and hands and the highlighted portions of their garments and contrasts beautifully with the warm dark brown of the paper and the cold grays, blues, dark greens, and umbers of the woods and shadows around them.

The numerals of the inscribed date on the field sketch are nearly indecipherable. April 10 may be most likely because of the landscape at that camp, but this may be immaterial since Möllhausen was mainly interested in illustrating the character of the high table-land. In his entry dated "Camp 71, Pine forest, April 10," Lieutenant Ives described how they left the "cedar forest" in the vicinity of their April 5 camp and travelled in a northeasterly direction. Near their camp for April 6, they "entered the region of pines. The growth was thicker, and trees of considerable size began to be mingled with the low cedars. The ascent from the Hualpais spring, though gradual, had been rapid, and the barometer indicated an altitude of about six thousand feet. The increase of elevation was felt very sensibly in the changed temperature, which had become wintry and raw."[1]

Möllhausen noted the presence of snowbanks, "the last remnants of larger accumulations," and specifically identified the tall pines as "Rocky Mountain white pine" or "Pinus flexilis, James":

> This beautiful symmetrical tree only grew here in a few places, but for us it was exceedingly pleasant to see, for since leaving California we had seen hardly anything but deserts with only a few trees. These pines towered as high as eighty feet above their roots. Even the cedars were shaped more like trees. The cedars and a low-growing variety of pine [identified as "Pinus edulis" in a footnote] formed dense uniform groves spread over broad areas.[2]

40B

John James Young after Balduin Möllhausen, *Camp — Colorado Plateau*, toned lithograph (black with gray ruled screen), from Ives, *Report* (1861), part 1, opp. p. 107, Amon Carter Museum Library

40A

Balduin Möllhausen, *Landscape on the High Plateau, April 10th [Landschaft, 13. April]*, graphite on paper, from Sketchbook 3, Möllhausen Family, Bleicherode, Germany, photograph courtesy of Dr. Friedrich Schegk, Munich

Approximately an inch of snow fell on the night of April 6. A soldier who had gone out hunting lost his way and was missing for two days, causing them to remain an extra day in that camp. By April 8 all of their Indian guides had deserted them, but the men continued to follow a steep Indian trail toward the northeast, roughly paralleling modern Arizona State Highway 18 in Coconino County. The weather continued to worsen, causing the men and animals grievous suffering from the cold, and that afternoon a full-blown snowstorm accompanied by lightning and thunder forced them to halt near the modern site of Frazier Wells.[3] Ives wrote:

> The day was nearly spent; the packs were therefore taken off, camp made, fires kindled, and the mules driven into a ravine. About sunset it promised to clear off, but the clouds reassembled, the wind and sleet again drove past, and the night was bleak and raw. The unfortunate mules, benumbed with cold, stood shuddering about the fires that were made in the ravine. The sudden change from hot summer weather was a severe test of endurance, and there was danger that in their weak condi-

tion they would not be able to stand it. The snow and the gale continued nearly all of the next day. The grass was entirely covered. The animals had to fast for twenty-four hours longer, and I thought that last night would have finished the majority of them, but singularly enough not one has died.[4]

Möllhausen described this campsite, where bad weather on April 9 confined them to their tents and they "spent the greater part of the day cleaning . . . rusty weapons and putting them in order." When the sun came out the next day, he "sat in the top of a high tree and revelled in the magnificent spectacle," watching as the silhouettes of distant trees "gradually faded one into the other":

> The lively activity close at hand stood in glaring contrast to the endless wilderness and its characteristic awesome loneliness. A dozen fires blazed merrily. Robust figures with bearded faces wearing shabby tattered clothing stood around them. Some dragged up dry wood. . . . [P]eavish ravens squawked from the bare branches of a dead fir, waiting impatiently for our departure so that they could search for tasty morsels in our deserted camp site. I climbed down from the lofty perch where I had enriched my memories with some beautiful scenes, and soon thereafter lay among my comrades on fragrant cedar branches beside a crackling fire.[5]

The expedition finally departed the camp early on April 11, still following the Indian trail through similar surroundings.

The wintry darkness conveyed by Möllhausen's watercolor is completely absent in John James Young's lithograph for Ives' *Report* (cat. 40b). Although most elements of the composition remain, they are altered: Young reduced the firewood collectors in the left portion of the composition to an absurdly tiny scale that distorts the perspective; in addition, all figures wear Young's characteristic tall, pointed-crown hats, and the tent is no longer pyramidal.

---

1 Ives, *Report*, part 1, pp. 103-104. Coues, ed., noted in *On the Trail of . . . Garcés*, pp. 335 n.19 and 409 n.18, that Ives' camp 71, "Pine forest," later became known as "Pine Spring." Garcés probably visited it in June and July 1776, calling it "Pozo de la Rosa" (Rose Well).
2 Möllhausen, *Reisen*, chapter 23.
3 Möllhausen, *Reisen*, chapters 23 and 24; Ives, *Report*, part 1, pp. 103-104; Egloffstein's map 2; and USGS 1:250,000 Map "Williams, Arizona."
4 Ives, *Report*, part 1, p. 104.
5 Möllhausen, *Reisen*, chapter 24.

Cat. No. 41

Balduin Möllhausen
***Cañon Where Ives Went Down/ View From the Hieght [sic]
(Precipice Leading to Cataract Cañon)***
Watercolor and gouache on paper
8⅝ x 11¼ in. (22.0 x 28.6 cm)
signed l.r.: "Möllhausen"
l.c. on paper mount: "Cañon where Ives went down./
View from the hieght [*sic*]"
ACM 1988.1.41

Cat. No. 41

On April 12, still hoping to find a route across the Colorado River through the Grand Canyon, Ives and his men continued northeast over the dry Coconino Plateau, along a route roughly equivalent to modern Arizona State Highway 18.[1] They had already travelled miles without finding sufficient water for their mules when they reached the head of a ravine and followed a "well-beaten Indian trail" into it. The ravine went deeper and deeper until, Ives wrote, "the bluffs on either side of it assumed stupendous proportions."[2] This was near the head of what is now known as Hualapai Canyon, a side canyon that meets Havasu Canyon (the modern name for the Colorado River end of Cataract Canyon) at Supai village.[3]

According to Ives, the trail along the right-hand bluffs of Hualapai Canyon,

> though narrow and dizzy, had been cut with some care into the surface of the cliff, and afforded a foothold level and broad enough both for men and animals. I rode upon it first, and the rest of the party and the train followed—one by one— looking very much like a row of insects crawling upon the side of a building. We proceeded for nearly a mile along this singular pathway; which preserved its horizontal direction. The bottom of the cañon had meanwhile been rapidly descending, and there were two or three falls where it dropped a hundred feet at a time, thus greatly increasing the depth of the chasm.[4]

Möllhausen later produced a watercolor of Hualapai Canyon and, like Ives, left a lengthy account of their trip into it:

The thought of having to continue along a trail suitable only for antelopes and mountain goats was a dizzying prospect. We dismounted in order to lead our animals by the reins. We unbuckled our spurs and began the descent in a long procession. The ravine opened up as it became deeper, and the opposing bluff gave us some idea of the one we were on. Each consisted of a row of immense, uniformly spaced towers of rock with high capped embayments. These led up at an angle to the cap rock on the plateau. A single vertical wall extended down into the horrible depths from the layer of rock which formed our trail. Its continuation was very prominent on the opposite side.[5]

He noted that the party's lively conversation ceased as the trail suddenly grew worse, and Ives reported that some "of the men became so giddy that they were obliged to creep upon their hands

41B
Cataract Canyon, March 1990

41A
John James Young after Balduin Möllhausen?, *Precipice Leading to Cataract Cañon*, engraving, from Ives, *Report* (1861), part 1, p. 106, Amon Carter Museum Library

and knees, being unable to walk or stand. In some places there was barely room to walk, and a slight deviation in a step would have precipitated one into the frightful abyss."[6] Eventually the trail became completely impassable for the mules, and the men themselves feared they would have to use ropes. Finding a place to turn the mules, Ives and his men retraced their steps several miles back up the canyon trail to the plateau. The lieutenant had the Mexican *arrieros* drive the mules back along the plateau to the last water they had seen—in lagoons some thirty miles distant—and ordered them to remain there for a couple of days while the rest of the men explored on foot. The latter group camped for the night near one of the entrances to the canyon.[7]

Early on the morning of April 13, Lieutenant Ives, Dr. Newberry, Egloffstein, Peacock, Lieutenant Tipton, Möllhausen, and six soldiers headed back down the trail to explore Hualapai Canyon and its branches. Möllhausen and Dr. Newberry separated themselves from the rest of the group to follow the canyon's course to the south, hoping, in Möllhausen's words, "to follow the canyon all the way from its beginning." Eventually finding their path blocked by precipices, he and Dr. Newberry retraced their steps. Since Lieutenant Ives and the rest of the group had proceeded farther down the canyon in the other direction, the two men decided to work their way up to the heights "to obtain an overview of the immediate vicinity, which in a geological and topographical sense concealed so much that was noteworthy and unusual." Making the difficult climb up the steep slopes to the top of the plateau, they strolled along the edge of Hualapai Canyon in a northwesterly direction. Möllhausen wrote:

> where the flimsy trail led far below us, we saw formations and scenery which we could not have thought of or imagined in our wildest fantasy. We could see down as far as 2000 feet to dark, red sandstone which formed the floor of the dry, barren, rocky canyon, and which continued to sink and open up in a westerly direction. The countless gullies resembled delicate veins as they meandered from the base of the vertical walls toward the center. There they united to form a deep river bed, which as far as the eye could see was the color of glowing red iron. Colossal embayments, capped evenly and symmetrically, jutted up out of the canyon. They had been formed from the horizontal layers of various epochs, and during thousands of years had been chiseled out by the effects of the atmosphere to a greater or lesser degree depending upon their degree of hardness. Ribbons of vivid colors extended along these carved walls in the order in which they had been laid down, and while Dr. Newberry attempted to decipher the geological history of this mighty tableland by carefully analyzing those layers, I took my sketchbook in hand and created a keepsake of this remarkable point.[8]

Möllhausen made this sketch around noon, in the "searing heat" of the sun, and recorded the "indescribable variations" of the lines and colors and the "dreadful desolation" of the landscape. Before they got up to leave, he also noted such peculiar details as "the faint somnolent buzzing of insects," and "numerous lizards" that "lay motionless on the warming rocks."[9]

Möllhausen's sketch apparently served as his model for the watercolor inscribed *Cañon where Ives went Down*. The view, taken from near the top of the plateau, on the eastern side of Hualapai Canyon above the present settlement of Hualapai Hilltop, looks

west-northwest across the canyon. Rising in the far distance are probably the Uinkaret Mountains.[10] He had difficulty conveying the canyon's vast dimensions but attempted to do so by introducing foreground figures: himself, with a rifle slung over his shoulder, sketching on a ledge in the left foreground, and Dr. Newberry, raising his geologist's hammer to chip away a specimen. The engraver for Ives' *Report* omitted the two explorers, rearranged some of the formations, and exaggerated the verticality of the distant canyon walls (cat. 41a). A recent photograph of Hualapai Canyon taken from approximately the same vantage point (cat. 41b) demonstrates that Möllhausen also simplified the formations considerably.

---

1 USGS 1:250,000 maps "Williams."; Ives, *Report*, part 1, pp.104-105.

2 Ives, *Report*, part 1, pp. 104-105. Much earlier, in June 1776, Padre Francisco Garcés was probably the first white man to follow this trail to the Supai village. See Coues, ed., *On the Trail of . . . Garcés*, pp. 335 n.20, 336 n.21. In November 1884, General Crook and several men, including John Gregory Bourke, followed this same trail, known as the "Hualpai Trail," out of the canyon after having visited the Indians at Supai. See Frank E. Casanova, ed., "General Crook Visits the Supais as Reported by John G. Bourke," *Arizona and the West* 10 (Autumn 1968): 253-276.

3 Compare USGS 1:250,000 map "Grand Canyon" and Egloffstein's map 2. According to Barnes, *Arizona Place Names*, p. 202, "Coues says Padre Garces called this creek and canyon 'Rio Jabesua' which he thinks was merely Garces' way of spelling their name. He sometimes called the Indians 'Agua Azul' (Blue Water) for the color of their water in the springs. After the Havasu or Supai Indians." See note on the Havasupai Indians in cat. no. 42. This canyon is part of the Havasupai Reservation, created on June 8, 1880, under the administration of President Rutherford B. Hayes.

4 Ives, *Report*, part 1, pp. 105-106.

5 Möllhausen, *Reisen*, chapter 24.

6 Ives, *Report*, part 1, p. 106. See also Möllhausen, *Reisen*, chapter 24.

7 Ibid.

8 Möllhausen, *Reisen*, chapter 25.

9 Ibid.

10 See USGS 1:250,000 map "Grand Canyon."

Cat. No. 42

Balduin Möllhausen
Doctor's Rest
Watercolor and gouache on paper
8½ x 11⅛ in. (21.6 x 28.3 cm)
Signed l.l.: "Möllhausen"
l.l. on paper mount: "4"
l.c. on paper mount: "Doctors rest."
verso, center on paper mount: "Doctors Rest"
ACM 1988.1.4

Cat. No. 42

One of Möllhausen's finest and most dramatic views of the Grand Canyon area is his watercolor inscribed "Doctor's Rest." He made the original field sketch for the watercolor on the heights approximately three quarters of a mile north of the present Hualapai Hilltop turnaround on the Havasupai Indian Reservation. The date was April 13, the same day that Lieutenant Ives, Egloffstcin, Pcacock, Lieutenant Tipton, and six soldiers proceeded deep into Hualapai Canyon, heading toward Havasu Canyon and Supai village, near the heart of the Grand Canyon (see essay, p. 51, and cat. no. 41).

Möllhausen and Dr. Newberry "walked along the rim of the plateau"—perhaps only a few yards from the earlier vantage point where he made his view of the *Cañon Where Ives Went Down*—and kept their eyes constantly

on the canyon, which became more extensive. It finally converged into a wide basin, into which new fissures emptied from all directions. . . . Lieutenant Ives and the rest of our companions reached this point in their seemingly subterranean passage. Standing up on the heights, we were offered an excellent, and almost the only, vista of this inaccessible region.

Later I showed my sketch to Lieutenant Ives and Mr. Egloffstein, but neither recognized the basin, for during their excursion the towering rock walls had constantly restricted their field of vision, and the deep canyons in which they found

42A
After Balduin Möllhausen?, *Side Cañons of the Flax River*, engraving, from Ives, *Report*, part 1, p. 109, Amon Carter Museum Library

42B
The junction of Hualapai and Havasu Canyons, with Mount Sinyala in the distance, August 1990

themselves looked to me from up above only as insignificant gullies. The clear atmosphere peculiar to those regions often causes errors when estimating distances, but I consider that my estimates are close to accurate when I calculate the breadth of this formidable rocky amphitheater at from six to seven miles. The bluffs exhibited exactly the same strata and colors which we had observed in the first canyon, except that the most singular formations adorned the colossal walls, or had been partially eroded away from them, and seemed to vie with one another in the most fantastic lines and forms. Red sandstone more than 2000 feet below formed the basin's apparently level surface. Many colored streaks eroded by water covered the tile-red expanse and these led us to think that new fissures and ravines must open up, leading thousands of feet farther down.

The last remnant of the plateau rose up in the center of the basin. Its even proportions stood in marked contrast to its jagged and craggy surroundings. An enormous butte jutted up from the red sub-stratum, and its highest point was adorned by a tower. Its rounded top had once been connected to the rock layer upon which we stood, and looking down on the tower and the butte's slopes, we recognized the horizontal strata everywhere. We could easily make out their continuation on every vertical cross-section of the elevated tableland. Colossal masses were compressed into a single image in this vast space, and yet were only a fraction of an immense whole. The mighty dimensions of the individual walls and towers seemed to fade in comparison to those in the narrower canyons, yet one almost shuddered at the sensations which the colorful, intricately formed, rigid, motionless masses exercised. One hardly dares to raise his voice when standing face to face with such terribly beautiful nature.[1]

A photograph (cat. 42b) taken from the promontory on almost the exact spot where Möllhausen and Dr. Newberry must have stood proves the general accuracy of the watercolor. All of the topographical features appear to line up correctly, although the artist brought some of them forward for emphasis. Hualapai Canyon continues up to the middle distance of the sketch, where it joins Havasu Canyon. The symmetrical butte rising in the center of the basin is Mount Sinyala, 5,434 feet high and situated on Sinyala Mesa. The crevice running from left to right in front of the base of the butte represents the deep canyon formed by Havasu Creek, while another huge crevice beyond and to the left of the butte apparently denotes the course of the mighty Colorado River and part of Sinyala Canyon. In the far distance to the north rises the Kanab Plateau, which forms the north rim of the Grand Canyon.[2] The two figures in the left foreground of the watercolor represent Dr. Newberry and the artist himself. Fine touches include a bald eagle soaring above the chasm in the center foreground and dry-brush highlights on the plateau at right.[3]

An engraving in Ives' *Report* titled *Side Cañons of Flax River* (cat. 42a) exhibits a remarkable similarity to Möllhausen's watercolor of *Doctor's Rest*. It clearly depicts Mount Sinyala and the great basin formed where Havasu and Hualapai Canyons merge near the Colorado, but the vantage point has been altered, providing a clearer view of one of the deeper gorges. Although the print could be based on another sketch by either Möllhausen or Egloffstein, the engraver may simply have taken Möllhausen's *Doctor's Rest* sketch and invented or rearranged some of the foreground formations.[4] The title of the engraving reflects their mistaken belief that these canyons branched off from the "Flax" or Little Colorado River.[5]

None of the written sources specifically mention whether Egloffstein made sketches on his excursion with Lieutenant Ives to the bottom of Hualapai and Havasu Canyons on April 13 and 14. However, two prints after Egloffstein in Ives' *Report* suggest that he did. One print (cat. 42c), redrawn by J. J. Young, purportedly depicts a rock tower along the walls of the *Big Cañon*, though it probably is of Hualapai Canyon. The exaggerated depth and verticality of the view suggest that artist was deep within the canyon, yet the top of the opposite plateau can be seen from above. These distortions may be the work of Egloffstein, the engraver, or both.[6]

---

1 Möllhausen, *Reisen*, chapter 25.
2 USGS 1:250,000 map "Grand Canyon"; also compare USGS 15' series map "Kanab Point Quadrangle" (1962). Barnes notes in *Arizona Place Names*, p. 409: "Sinyala Mesa or Mountain, Coconino Co., Map, Grand Canyon N.P., 1927. Elevation 5,445 feet. Near lat. 36° 19', long. 112° 43'. Low tableland, left bank Colorado river, between Havasu creek and Sinyala canyon. Name suggested by Charles Sheldon for Havasupai chief. Decision U.S.G.B."

42c
John James Young after F. W. von Egloffstein, *Big Cañon*, transfer
lithograph, from Ives, *Report* (1861), part 1, opp. p. 110,
Amon Carter Museum Library

3 According to Bryan T. Brown, Steven W. Carothers, and R. Roy Johnson, *Grand Canyon Birds: Historical Notes, Natural History, and Ecology* (Tucson: The University of Arizona Press, 1987), pp. 169-170, the bald eagle "is seen throughout the year in Grand Canyon as a rare transient. . . . During migration, it is seen soaring along the rims on the thermal updrafts. The majority of bald eagles are seen along the river in winter."

4 It should be noted that in our recent examinations of the area, time and the vast distances around the basin made it impossible to confirm all of the possible angles from which the view might have been taken.

5 Barnes accounts for some of the confusion in *Arizona Place Names*, p. 164: "Flax River—'Rio De Lino.' Sp., 'flax, linen.' One name for the Little Colorado river. Ives called it this for the plant that grew on its banks. Newberry always referred to it as 'Little Colorado River.' His report, written in 1861, may explain the change. Beale, in 1859, always called it the 'Colorado Chiquito.'"

6 The second Egloffstein print, a lithographic panorama titled *Cañon of Cataract Creek and Flax River*, in Ives, *Report,* between pp. 122 and 123, has an unusual composition. A rocky pinnacle rising in the center separates two vistas that open up at left and right. This central obstruction probably indicates that the artist was well below the top of the plateau when he made the original view, although the lack of shading or tonalities in the lithograph make it difficult to recognize or identify the exact terrain. Underneath the righthand vista is a key reading "Cataract Creek," but without the original sketches for both prints, many observations remain speculative.

CAT. NO. 43

Balduin Möllhausen
Cañon near Upper Cataract Creek (Side Cañons of the Colorado)
Watercolor and gouache on paper
8½ x 11¼ in. (21.6 x 28.6 cm)
signed l.r.: "Möllhausen"
l.l. on paper mount: "44"
l.r. on paper mount: "Cañon near upper Cataract Creek"
ACM 1988.1.42

Cat. No. 43

Möllhausen's dramatic water-color titled *Cañon Near Upper Cataract Creek* looks northeast from the head of Driftwood Canyon into Cataract Canyon and shows a section of the Grand Canyon in the far distance. Recent photographs (cat. 43b) taken at or near the spot where the artist first sketched the scene attest to the degree of verisimilitude he attained.[1] Möllhausen emphasized certain formations by contrasting light and shade, just as in real life the constantly changing positions of clouds and the sun create various effects upon the distant rocks. The two shadowy foreground figures (perhaps Möllhausen, Dr. Newberry, or Egloffstein) and the scrub cedar bushes act as ré-poussoir elements that frame the composition and fix the scale.

Ives alluded to this landscape in his entry dated April 14, from their camp on the Colorado plateau:

The region east of camp has been examined to-day. The extent and magnitude of the system of cañons in that direction is astounding. The plateau is cut into shreds by these gigantic chasms, and resembles a vast ruin. Belts of country miles in width have been swept away, leaving only isolated mountains standing in the gap. Fissures so profound that the eye cannot penetrate their depths are separated by walls whose thickness one can almost span, and slender spires that seem tottering upon their bases shoot up thousands of feet from the vaults below.[2]

Möllhausen also wrote about this area in his usual verbose manner, and his details augment his watercolor and several prints, including one by Egloffstein. His other sketch of the "rocky amphitheater" (the basis for *Doctor's Rest*, cat. no. 42) having "aroused much interest" among those who had just returned that morning from their hike in the canyon floor the day before,

Egloffstein wanted to enjoy such a scene, and despite his sore feet he decided to accompany Dr. Newberry and me on a new excursion in the afternoon. This time we selected a more northerly [easterly] direction, for we had discovered a deeper depression in the ground which might possibly be the deep-lying bed of the Little Colorado. Our immediate goal was to locate it again, for we had given up all hope of getting down to the Great Colorado at that latitude.[3]

Reaching "the rim of the canyon after a three-mile march," they encountered a vista similar to the earlier view shown in *Doctor's*

43A
John James Young after Balduin Möllhausen?, *Side Cañons of the Colorado*, engraving, from Ives, *Report* (1861), part 1, p. 101, Amon Carter Museum Library

Rest but quite "different in its individual parts and forms." As Möllhausen explained:

> The impression which the rocky chasm made upon us was intensified because we were standing right on the edge of the plateau and the dreadful abyss dropped off almost beneath our feet. We looked down with trepidation at the dark-red floor of the arid basin, which lay some 2000 feet below us. The various gullies were incised in countless sinuosities resembling arabesque filigree. They merged with the canyons, which extended far down into the basin from deep fissures in the plateau. The average breadth of this rocky cauldron was at least six miles. The chasm was divided into two halves by an extension of the plateau which resembled a great wall, and it was adorned with such peculiar formations that it actually looked like the well preserved ruins of an Indian city. A colossal amphitheater which extended out in an evenly proportioned arc from our vantage point to the rocky wall crowned with ruins was even more noteworthy. It was connected to the main gorge by a wide aperture, but as a complete formation it was more overpowering than any of the other sights.[4]

Möllhausen even described how he

> sat once again on a dizzying promontory on the edge of the abyss and sketched. Formations from the various epochs rose up one upon the other out of the depths. They could easily be recognized by their starkly contrasting colors, with each layer representing one of the ages. The walls stood perpendicular, as if the slightest tremor might cause them to come tumbling down, and as a reminder of eternity, I could see indications which clearly demonstrated that falling drops of water had formed the gorges which towered up all around me. I . . . looked with longing over to the high rock ramparts which rose up out of gorge about 20 miles away, for along its base either the Big or Little Colorado River probably flowed.[5]

The men speculated that the two rivers had gradually eroded the plateau from above to create the huge canyons rather than undercutting them by subterranean paths. Möllhausen also noted that "the open cleft through which the two rivers presumably flow" was "visible for a great distance from the heights of the San Francisco Mountains." He remarked that he "gazed across at the immense canyon wall" with "a certain melancholy" and watched "a kite soaring over the depths on steady wings at the same elevation" as his own viewpoint. Then he attempted a prophetic statement that turned out to be way off the mark:

> With awe I imagined a picture of the rocky canyon of the "Colorado of the West," which perhaps for coming centuries will remain a secret to mankind. As I turned around in order to return to camp, I once again had the apparently unbroken plain before me. The heavens had become overcast. A few rose-colored streaks shimmered in the west, announcing the impending sunset, and I hastened so I would not be caught by surprise in the canyons after dark.[6]

They returned to camp to learn that two soldiers who had gotten lost as they returned from the canyons had been found. Later in the evening the Mexican *arrieros* arrived at camp with the mules. They prepared to head south the next morning (April 15), since other side canyons prevented them from travelling eastward. Ives hoped to go around these barriers, cross the Little Colorado, "and again travel north upon the opposite side of that stream."[7]

The engraver for Ives' *Report* (cat. 43a), working from *Cañon near Upper Cataract Creek* and an undoubtedly similar watercolor titled *Side Canyons of the Colorado* which Möllhausen gave the Berlin

43B
Upper Cataract Creek, March 1990

43c
After Balduin Möllhausen, *Schluchten im Hoch-Plateau und Aussicht auf das Colorado Cañon*, toned electrotype engraving or transfer lithograph, from Möllhausen, *Reisen* (Leipzig: Otto Purfürst, 1861), vol. 2, opp. p. 100

The same formations appearing in Möllhausen's watercolor are found here, but once again the printmaker(s) reinterpreted them. The foreground is nearly unrecognizable; the view may have been made from a different vantage point, but more likely this change reflects the printmaker's artistic license.

Ives and his men retraced their tracks to the forest lagoons,[10] where they camped the nights of April 16, 17, and 18. Meanwhile, Möllhausen reported:

> Our indefatigable Egloffstein set out on foot toward the Colorado to learn as much as possible about the geographical location of this stream, while I roamed through the adjacent [near-by] vicinity with my bird rifle to learn a little bit about the migrating birds which had alighted on the plateau. . . . Egloffstein did not return from his tiring excursion until late in the evening [of April 17]. He had had little success.[11]

As a result of Egloffstein's trek, however, Ives was able to report:

> An excellent view was had of the Big cañon. The barometric observations upon the surface of the plateau, and at the

Geographical Society in 1904,[8] took considerable liberties with the topography by compacting the composition and emphasizing the deeper ravines and fissures. As in other prints produced by the American printmakers, Möllhausen's original foreground details are almost ignored: the two foreground figures are reduced in scale and placed upon dangerous and unlikely precipices, and the character of the foreground rocks has been altered.[9] A similar print titled *Schluchten im Hoch Plateau und Aussicht auf das Colorado-Cañon* ("Side Canyons in the High Plateau and View of the Colorado Canyon"; cat. 43c) appeared in *Reisen*, but the German printmaker, Leutemann, carefully reproduced Möllhausen's topography. Since Leutemann was primarily an animal painter, he added dramatic details in the foreground to introduce a narrative element to the composition: a young lamb bleats over its slain mother while a bald eagle alights on a dead tree and wolves or coyotes move in for the spoils.

The sketch (or sketches) that Egloffstein made on April 14 are now lost, but one apparently was the basis for J. J. Young's print titled *Upper Cataract Creek, near Big Cañon* (cat. 43d) in Ives' *Report*.

43d
John James Young after F. W. von Egloffstein, *Upper Cataract Creek, near Big Cañon*, transfer lithograph, from Ives, *Report* (1861), part 1, opp. p. 108, Amon Carter Museum Library

mouths of Diamond and Cataract rivers, showed that the walls of this portion of the cañon were over a mile high. The formation of the ground was such that the eye could not follow them the whole distance to the bottom, but as far down as they could be traced they appeared almost vertical. A sketch taken upon the spot by Mr. Egloffstein does better justice than any description can do to the marvelous scene.[12]

In a very short-sighted way, Ives concluded:

Our reconnoitering parties have now been sent out in all directions, and everywhere have been headed off by impassable obstacles. The positions of the main water-courses have been determined with considerable accuracy. The region last explored is, of course, altogether valueless. It can be approached only from the south, and after entering it there is nothing to do but leave. Ours has been the first, and will doubtless be the last, party of whites to visit this profitless locality. It seems intended by nature that the Colorado river, along the greater portion of its lonely and majestic way, shall be forever unvisited and undisturbed. The handful of Indians that inhabit the sequestered retreats where we discovered them have probably remained in the same condition, and of the same number, for centuries. The country could not support a large population, and by some provision of nature they have ceased to multiply. The deer, the antelope, the birds, even the smaller reptiles, all of which frequent the adjacent territory, have deserted this uninhabitable district.[13]

5 Ibid.
6 Ibid.
7 Ives, *Report*, part 1, p. 109, and Möllhausen, *Reisen*, chapter 25.
8 See the 1904 list, number 45 (*Seitenschluchten des Colorado*); this watercolor was among the ones lost or destroyed in World War II.
9 An almost exact copy of the American print, titled *The Chasm*, appeared in Alpheus Hyatt, "The Chasms of the Colorado," *American Naturalist* 2 (September 1868), opp. p. 364, plate 8.
10 On their way they decided to follow the mysterious wagon trail that crossed their path. They found that it led northwest and then turned around, having been barred from further progress by the vast canyonlands. Subsequent investigations confirmed their suspicions that the false trail had been blazed by Lieutenant Beale. Ives, *Report*, part 1, p. 110; Möllhausen, *Reisen*, chapter 25.
11 Möllhausen, *Reisen*, chapter 25.
12 Ives, *Report*, part 1, p. 110. This sketch was reproduced as a lithographic panorama titled *Big Cañon from Colorado Plateau* in Ives' *Report*, between pp. 112 and 113.
13 Ives, *Report*, part 1, p. 110.

1 Compare with USGS 1:250,000 map "Grand Canyon"; USGS 15 min. series map "Havasu Point Quadrangle" (1962).
     Barnes, *Arizona Place Names*, p. 83, notes: "Cataract, Canyon and Creek, Coconino Co., U.S.G.S. Map, 1923; G.C.N.P., 1927. Heads at Williams, flows northwest into Grand Canyon. Named for a series of beautiful waterfalls. See Elliott's *History of Arizona*. Garces, 1776, called it 'Rio de San Antonio.' Ives, 1858, called it 'Cataract' and also 'Cascade Creek.' The Canyon begins about 20 miles northeast of Williams, runs northwest and enters Grand Canyon in Tusayan N.F. near line between Rs. 4, 5 W. Lower end of canyon for some 10 or 12 miles is called Supai or Havasupai canyon."
2 Ives, *Report*, part 1, p. 109.
3 Möllhausen, *Reisen*, chapter 25.
4 Ibid.

Cat. No. 44

Balduin Möllhausen
Bill Williams' Mountain
Watercolor and gouache on paper
8½ x 11½ in. (21.6 x 29.2 cm)
signed l.r.: "Möllhausen"
l.l. on paper mount: "16"
l.c. on paper mount: "Bill W. M."
verso, center, on paper mount: "Bill W. M."
ACM 1988.1.16

Located in Coconino County, Arizona, about six miles southeast of the modern town of Williams, is 9300-foot Bill Williams Mountain, named for the scout, guide, and hunter William Sherley Williams; its name first appeared on the map by artist and topographer Richard Kern for Sitgreaves' 1851 expedition.[1] Möllhausen and Ives first saw the mountain in December 1853 and early January 1854, as they travelled from east to west with the Whipple expedition.[2]

On April 11, 1858, as the Ives expedition headed north-northeast across the Coconino Plateau toward the canyonlands, Möllhausen recognized San Francisco and Bill Williams Mountains "over eighty miles" away, their picturesque peaks standing "in sharp relief, discernible by the white and pale blue areas and lines."[3] The mountains remained visible intermittently as long as they travelled along the plateau. On April 23, viewing them from near Partridge Creek, Möllhausen commented on the unusual warmth of the day, which "seemed to be warm up to the higher atmosphere, because even the summits of Bill Williams Mountain and the higher slopes of the San Francisco Mountains gradually emerged from their white cloak and shimmered with the pale color of winter grasses where they were not covered with tall pine forests."[4]

The expedition members noted the incredible change in the character of the landscape on April 25. Riding east toward the base of Bill Williams Mountain, Möllhausen observed:

> this virtually trackless wilderness seemed like an enticing garden, with its few but varied trees, its hills and dales, its rocky and fertile soil. All these formed very pleasant and surprisingly beautiful images and vistas. . . .

Tall pines soon predominated as we climbed up through this beautiful landscape. Beyond it to the north and northeast, we could see round, volcanic hills, while to the east we had the scenic panorama of Mount Sitgreaves and the San Francisco Mountains.[5]

On the same day Lieutenant Ives reported:

> This morning we re-entered the region of the pines, and have travelled all day in the midst of picturesque and charming scenery. The valleys are covered with a bright green sward, and open groves are disposed gracefully upon the lowlands and ridges. Heavy masses of snow are still piled upon the San Francisco summit, and this close proximity of winter heightens and gives a zest to the enjoyment of spring.[6]

Crossing Whipple's trail and picking up Beale's wagon road, they soon were "far to the south . . . on the edge of a prairie" two miles from Bill Williams Mountain. There,

44A
John James Young after Balduin Möllhausen, *Bill Williams's Mountain,*
engraving, from Ives, *Report* (1861), part 1, p. 114,
Amon Carter Museum Library

[t]he prairie was partially inundated by water from the mountain, and many meandering streams and washes cut across it. The melted snow flowed rapidly through these streams and washes in a wide arc to join the western tributaries of Bill Williams Fork or the Gila. We had made ten miles since morning, and since we found good grass, wood and water on the edge of the prairie, we decided to spend the night there, quickly making camp.[7]

Möllhausen's watercolor portrays this camp (no. 80), in a meadow at the mountain's base. Ives further described the area:

Bill Williams's mountain . . . though the second in importance of the cluster, is far less lofty than its colossal neighbor [San Francisco Mountain], and the snows that whitened its crest a few weeks since have nearly disappeared. A sparkling brook now dashes down the ravine and meanders through the centre of the meadow, which contains perhaps five hundred acres, and is covered with a luxuriant growth of grama grass. Statcly pines and spruce are scattered upon the surrounding slopes, and afford a delightful shade. We found in possession of the spot a herd of antelope that scoured over the mountain like the wind when they saw the train approaching.

To eyes that have been resting upon the deserted and ghastly region northward this country appears like a paradise. We see it to the greatest advantage. The melting snows have covered it into a well-watered garden, and covered it with green meadows and spring flowers. The grass, even when dried by the summer's sun, will remain nutritious. The groves of trees will at all times give the region a habitable appearance, and though it is not known how great the supply of water would be during the summer, the country can never present the arid wastes that are spread along the belts of territory both north and south.[8]

The engraving for Ives' *Report* (cat. 44a) reproduces the topography of Möllhausen's watercolor with some fidelity, but it alters some of Möllhausen's interesting details, such as the kicking mule, the traces of snow, and the pyramidal shapes of the tents. Most importantly, however, the print fails to convey what interested the artist most about the actual scene, the light and coloring, which Möllhausen the artist effectively captured, and which Möllhausen

the writer also described:

At our camp near Bill Williams Mountain, it was the magnificent evening illumination with all its shadings and nuances which interested me most and which now comes back so vividly to mind. Who during his life had not admired beautiful mountain formations and somber pine forests which commonly occur on both continents? Thus on that evening I sat on a pile of lava rocks and turned my undivided attention to Bill Williams Mountain on which grassy clearings were so picturesquely interspersed with thick dark-colored conifers. Narrow ribbons of snow adorned the steep slopes, and a delicate mist caused by a temperature difference in the damp spring air, floated over the heavily wooded canyons. Dr. Newberry sat at my side, and our admiration turned to audible amazement as the sun sank behind the western mountains and a rosy light flooded over the mountains and valley. It was like the glowing in the Alps, only milder and more delicate. Without disturbing the hue of the trees, the clearings, or the last remnants of snow, this magical luster enveloped the smallest protruding objects, and long after dusk had settled over the bottom lands and black shadows had descended into the canyons, the mountain peaks still appeared as if they were artificially illuminated by a Bengal fire. We arose as the last rosy glow gave way to darkness. As we strolled back to camp we saw the wild bearded faces of the men of our expedition glowing with the flames from resinous wood fires.[9]

The artist gave another watercolor version of this subject— titled *Bill Williams Mountains*—to the Geographical Society in Berlin in 1904. It was lost or destroyed in World War II.[10]

---

1   This double-topped lava cone rises to 9341 feet directly south of Williams, Arizona. Williams, an eccentric mountain man, was called Old Bill before he was forty. He was born in North Carolina in 1787 but moved with his parents to Missouri (then Spanish territory) in 1795. Over the next thirty years, he was by turns an itinerant preacher, missionary to the Osage Indians, and a hunter and guide, probably trapping the Rockies one at a time. In 1825 a federal commission hired him as guide, hunter, and interpreter while they surveyed the road from Fort Osage, Missouri, to Santa Fe. Afterward he resided in Taos but trapped and traded throughout the Rocky Mountains until the fur trade declined. He apparently trapped the streams of Arizona and lived with the

Hopis between 1835 and 1845. Antoine Leroux, another veteran trapper, met him at the mouth of the river named for him in 1837. Williams guided J. C. Frémont's ill-fated expedition of 1848-49 and in the spring of 1849 was mistakenly killed by Ute Indians (his wife's kinsmen) while he was retrieving Frémont's baggage. In October 1851, Leroux led the Sitgreaves expedition to a camp south of the mountain, on a stream supposed to be the head of Bill Williams Fork. Richard Kern, artist-cartographer with Sitgreaves, memorialized Old Bill by naming the mountain for him. Whipple subsequently realized that Bill Williams Fork headed much farther west, but the name stuck to the mountain, and in 1882 builders of the A.T. & S.F. railroad named the town on the mountain's north slope for Old Bill. See Frederic E. Voelker, "William Sherley (Old Bill) Williams," *Mountain Men and Fur Trade of the Far West*, vol. 8 (Glendale, California: A. H. Clark Co., 1971), pp. 365-394; Granger, *Arizona's Names*, p. 65; Wallace, "Sitgreaves Expedition," p. 348; Weber, *Richard H. Kern*, pp. 168-169.

2 Whipple, "Report," in *Explorations and Surveys*, vol. 3, part 1, pp. 83, 84-86; part 2, pp. 32, 33; Möllhausen, *Diary of a Journey*, vol. 2, pp. 161, 163, 164, 171, 172; Foreman, ed., *A Pathfinder in the Southwest*, p. 174, 176-179.

3 Möllhausen, *Reisen*, chapter 24.

4 Möllhausen, *Reisen*, chapter 26.

5 Ibid.

6 Ives, *Report*, part 1, p. 113.

7 Möllhausen, *Reisen*, chapter 26.

8 Ives, *Report*, part 1, pp. 113-114.

9 Möllhausen, *Reisen*, chapter 26.

10 1904 checklist of watercolors to the Geographical Society of Berlin, no. 47.

CAT. NO. 45

Balduin Möllhausen
Tree [Alligator Bark Juniper (Juniperus deppeana)]
Watercolor and gouache on paper
10½ x 9 in. (26.7 x 22.9 cm)
signed l.r.: "Möllhausen"
l.l. on paper mount: "31"
l.r. on paper mount: "58"
ACM 1988.1.29

CAT. NO. 46

Balduin Möllhausen
Fremont [Pinyon Pine (Pinus edulis)]
Watercolor and gouache on paper
10⅝ x 9 in. (27.0 x 22.9 cm)
signed l.r.: "Möllhausen"
l.r. on paper mount: "30"
m.r. on paper mount: "Fremont."
ACM 1988.1.28

Möllhausen's duties included sketching botanical specimens which could not be transported back to Washington, such as these trees. He recorded no title or date on one of his finished watercolors of a tree (cat. no. 45), thus making full identification difficult. Nonetheless, his journal entry of April 24, written near Bill Williams Mountain, may possibly illuminate this particular sketch:

The most significant thing we happened upon was a new species of cedar, which differed from the ordinary Juniperus occidentalis through a taller and more slender growth, as well as by its unusual [peculiar] bark. The latter did not show the twisted fibrous bast, but was more like the scarred and repeatedly cracked bark of a hundred-year-old oak. In considering this, Dr. Newberry named it oak-bark cedar. Incidentally, in Capt. L. Sitgreaves' *Report of an Expedition down the Zuñi and*

Colorado rivers, I find an illustration of such a cedar under the title "Rough Barked Cedar (Juniperus pacaderma)," which undoubtedly is the same species.[1]

The tree to which Möllhausen and Sitgreaves (or rather his naturalist, Dr. Samuel W. Woodhouse) referred was the alligator juniper, or *Juniperus deppeana*.[2] First described in Mexico in 1826, the species inhabits dry, arid mountain slopes and ranges from West Texas to central Arizona and south to central Mexico.[3] It sometimes shows "great eccentricity of both trunk and limb," as may be seen in Möllhausen's sketch.[4]

The watercolor titled "Fremont" (cat. no. 46) possibly depicts a pinyon pine (*Pinus edulis*). The inscription probably refers to John C. Frémont, or more specifically, to the description of the species in John Torrey's *Plantae Frémontianae*, a catalogue of all things collected by Frémont.[5] *Pinus edulis* abounds on the Colorado plateau and has been described as a "tree, rarely thirty or forty feet in height, with a short often divided trunk occasionally two and a half feet in diameter, but usually much smaller, and often not more than twelve or fifteen feet tall. During its early years, when the branches are horizontal, it forms a broad-based compact pyramid, and in old age a dense low round-topped broad head."[6]

Möllhausen included figures in both of these watercolors, no doubt to add a sense of scale. The figures—one standing and one seated—could be self-portraits or could depict other expedition members.

In addition to his specimen-sketching duties, Möllhausen may have been inspired by German

Cat No. 45

Cat No. 46

romantic painting and poetry, which often attached profound symbolic meaning to solitary trees.[7] He sketched lone trees on several occasions, and these sketches exist in various collections.[8]

1 Möllhausen, *Reisen*, chapter 28. The Sitgreaves report lithograph, titled *Juniperus Plochyderma (Rough Barked Cedar), (Torrey) Camp 19*, was based on a drawing by Richard Kern.
2 I am grateful to Dr. Richard Heavily and Dr. Andrew Wallace of Northern Arizona University at Flagstaff, who confirmed the plausibility of this identification as well as that of cat. no. 46.
3 Princeton botanist John Torrey called it *Juniperus pachyphloea*, but taxonomists have since learned that it was first described in Mexico. See Elbert L. Little, Jr., "Older Names for Two Western Species of Juniperus," *Leaflets of Western Botany* 5 (November 1948): 125-132.
4 Charles Sprague Sargent, *The Sylva of North America* (Boston and New York: Houghton, Mifflin and Company, 1896), vol. 10, p. 86, noted that Indians also gather and eat the fruit, as do birds, bears, and other mammals. The wood is used for fence posts and fuel. See Frances Ernest Lloyd, "Two Junipers of the Southwest," *Plant World: A Magazine of Popular Botany* 9 (1906): 86-91.
5 John Torrey, *Plantae Frémontianae* (Washington, D.C.: Smithsonian Institution, and New York: G. P. Putnam & Co., 1853). Torrey also described this species in his article "Fremontia," in *Proceedings of the American Association for the Advancement of Science*, vol. 4 (1851), pp. 190-193, no. 1843. According to Sargent, *The Sylva of North America*, vol. 10, p. 57, "*Pinus edulis* was discovered in 1846 in the valley of the Rio Grande in New Mexico by Dr. F. A. Wislizenus." See also Ronald M. Lanner, *The Piñon Pine: A Natural and Cultural History* (Reno: University of Nevada Press, 1981), which details the history of the species, including the Frémont-Torrey connection, and also contains a cookbook of piñon nut recipes.
6 Sargent, *The Sylva of North America*, vol. 10, pp. 55-57. Also see Harry J. Baerg, *How to Know the Western Trees*, The Picture Key Nature Series, 2d ed. (Dubuque, Iowa: Wm. C. Brown Company, 1973), p. 29.
7 One is tempted to speculate that Möllhausen had in mind German romantic master Caspar David Friedrich's majestic canvas *The Solitary Tree (Village Landscape in the Morning Light; Harz Mountain Landscape)*, which was one of the foremost works (acquired in 1822) in the Nationalgalarie, Staatliche Museem Preussischer Kulturbesitz, Berlin.
8 For example, the Whipple Collection, Oklahoma Historical Society, has a beautiful sketch that Möllhausen made of an old cottonwood or "Alamo." The Neues Palais Collection, Potsdam, also has a finished watercolor by Möllhausen of a giant California redwood.

INDEX

*Note: **bold face** indicates illustration*

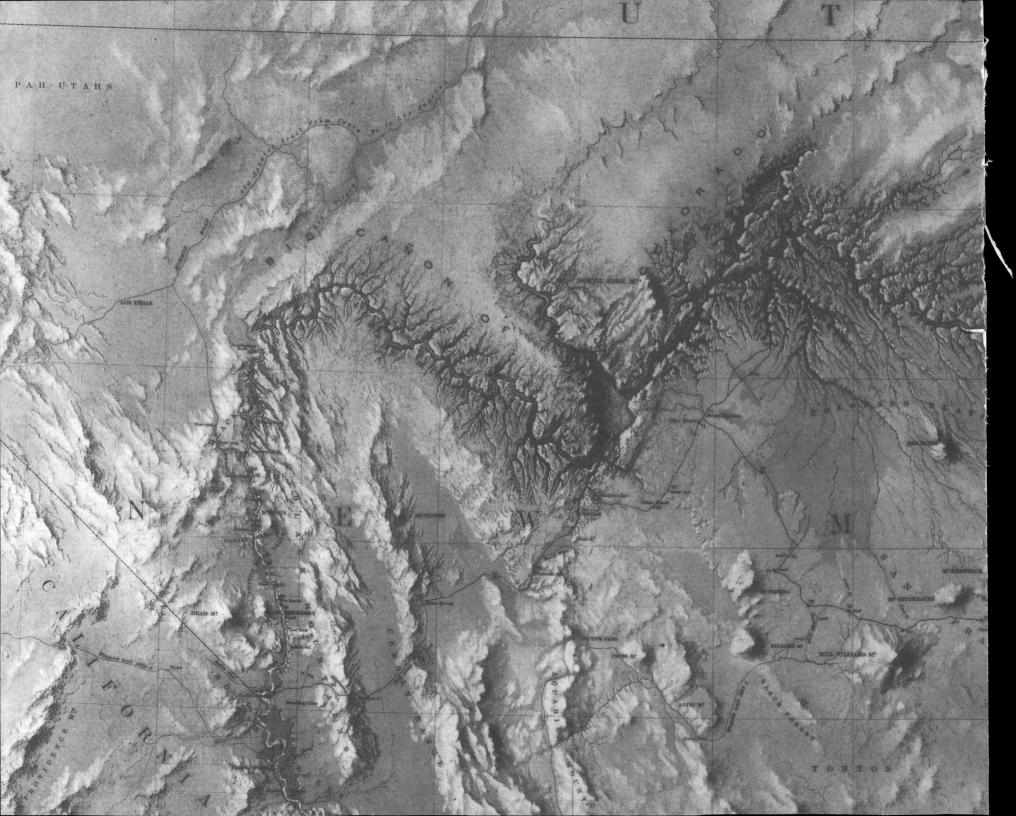